# ALWAYS READY, PERSISTENTLY UNDER-RESOURCED

---

## THE MODERN UNITED STATES COAST GUARD STORY

### AIRPOWER AND MARITIME FORCE MODERNIZATION SERIES
### BOOK 12

## ROBBIN LAIRD

Library of Congress Number: 2026900250

ISBN: 979-8-9939422-4-7

The cover image was created by an AI system.

Published by Second Line of Defense

Arlington, Virginia

# DEDICATION

Rear Admiral (Retired) Ed Gilbert inside a new
USCG maritime patrol aircraft during our visit to
the Miami air station in 2010.

This book is dedicated to Rear Admiral Edward "Ed" Gilbert,
United States Coast Guard (Retired), a leader, mentor, and friend

whose life of service embodies the best of the maritime spirit that has long defined America's guardians at sea.

Throughout his distinguished career, Ed Gilbert stood as an exemplar of integrity, quiet professionalism, and deep operational competence. He served not for recognition, but because he believed profoundly in what the Coast Guard represents: the unwavering commitment to protect lives, uphold the rule of law on the seas, and defend our nation's maritime interests wherever they may be challenged. His understanding that security, safety, and stewardship are inseparable dimensions of maritime power illuminated the Coast Guard's vital role in both peace and crisis.

Those privileged to serve with Ed remember a leader who commanded through respect rather than fear, and through example rather than rhetoric. He was known for listening before deciding, for asking hard questions before issuing easy orders, and for ensuring that the Coast Guard's operational excellence was always matched by moral clarity. To him, leadership was not about authority; it was about accountability, to his people, to his mission, and to the nation.

Ed's influence extended far beyond the decks and commands where he served. Many of those he mentored have gone on to lead in uniform, public service, and industry, each carrying forward his insistence on focus, competence, and enduring commitment to the maritime profession. His legacy rests not only in policy or precedent, but in people: the officers, enlisted members, and civilians who absorbed his quiet lessons in courage, discipline, and humility.

For me personally, Ed Gilbert was a compass, a fixed point amid the crosscurrents of change that mark the modern maritime domain. Our many conversations over the years, about leadership, joint operations, and the evolving role of the Coast Guard within the broader national defense enterprise, sharpened my understanding of American maritime power and its human foundation. He possessed a rare ability to connect the operational with the strategic, to link shipboard experience with policy realism, and to remind all who would listen that maritime security begins not with technology or budgets, but with people who are ready to sail into uncertainty and deliver.

The Coast Guard that Admiral Gilbert helped shape reflects his conviction that America's security must rest on agility, readiness, and the fusion of humanitarian purpose with strategic responsibility. Whether operating in polar ice, coastal waters, or disaster zones, Coast Guardsmen serve in that difficult space between defense and diplomacy, between crisis and calm. Ed understood this deeply and championed the cooperation across services, agencies, and nations that makes the modern Coast Guard indispensable to both national and global security.

In dedicating this book to Admiral Gilbert, I honor not only his service, but the broader generation of Coast Guard leaders who bridged the Cold War and post-9/11 eras, those who kept faith with the founding mission of the service even as its responsibilities expanded into new domains of cyber, counterterrorism, and global partnership. His career serves as a reminder that leadership at sea and ashore requires the fusion of tradition with transformation, heritage with innovation.

Rear Admiral Ed Gilbert's life exemplifies the Coast Guard's defining motto, Semper Paratus, "Always Ready." Ready to serve. Ready to lead. Ready to inspire. His enduring friendship, wisdom, and example have guided my work, not only in this book but in every effort to understand and convey the meaning of service to something larger than oneself.

This dedication stands as a small expression of gratitude and respect for a remarkable officer and friend whose life continues to guide those of us still charting our own courses in his wake.

# CONTENTS

# INTRODUCTION

Early on the morning of September 11th, 2001, I arrived at the Pentagon for what I thought would be a routine meeting about post-Soviet nuclear security issues. This was the work that had defined much of my career since the end of the Cold War, understanding security challenges in the former Soviet republics, trying to make sense of a world that had fundamentally shifted but still carried the deadly remnants of superpower competition. I parked just outside the Defense Secretary's office, and the morning felt ordinary, carrying the familiar rhythm of bureaucratic business.

As I sat waiting for my meeting, the television showed an airliner plowing into the World Trade Center. The atmosphere in the Pentagon remained remarkably normal, even as we watched the images from New York. There was little realization that the Pentagon might be next, that we were sitting in what would become the next target. The cognitive distance between what we were seeing on television and our immediate reality was profound.

When I asked one of the officials whether they were concerned about a similar event happening to the Pentagon, the person said they did not know if this was simply an accident. As a former New Yorker,

I was certain it was not. The idea that a commercial airliner would accidentally fly into the World Trade Center on a clear September morning defied all logic and experience.

Then, suddenly, I felt the building rock. Chaos erupted in the building as we rushed outside to see what was happening.

In that moment of confusion, I saw General Myers, Chairman of the Joint Chiefs, and Secretary of Defense Rumsfeld passing me, going back into the building. There was a determination on their faces that spoke to the immediate transition from peacetime routine to wartime crisis. I got into my car to drive home, and as traffic stopped on Interstate 395, I looked over to see the plane fitted inside the Pentagon. The aircraft was still intact at that point, much more of the plane survived the initial impact than was later reported. It was a surreal sight: a commercial aircraft embedded in one of the most secure buildings in America, the symbol of our military power violated in the most direct and brutal way imaginable.

For several days after the attack, we could smell the smoke and remains in our area of Arlington. That pungent smell will linger in my mind and heart forever. It was the smell of burning jet fuel, of destruction, of death. It hung over our neighborhood like a physical reminder we couldn't escape, even in our own homes. Daily life became surreal, with the Pentagon remaining both a crime scene and a military headquarters simultaneously trying to function.

## The Coast Guard Before the Storm

In the years leading up to September 11th, I had become acquainted with the United States Coast Guard through work with a maritime security company in the late 1990s. I focused on a challenge that would prove prescient: bad actors could access modern technology rapidly without going through complicated acquisition processes, tapping into the commercial world much more effectively than the U.S. Coast Guard could. This observation would become increasingly relevant as the security landscape transformed.

The Coast Guard was then developing what would become known as the Deepwater program, an ambitious attempt to address the block obsolescence of the Coast Guard fleet. The program represented an early and bold effort to create a multi-domain acquisition strategy, based on the idea that payloads and connectivity could deliver what the Coast Guard needed, not just the most expensive platforms. It was pioneering work in thinking about acquisition in multi-domain terms, a challenge we still face today, though it emerged in an era when such thinking was in its infancy.

The Coast Guard's mission set has always been extraordinarily broad, encompassing search and rescue, law enforcement, environmental protection, port security, and national defense. This breadth creates inherent tensions in resource allocation and strategic focus. The organization was and is characterized by incredibly dedicated men and women facing missions and challenges beyond their available resources, never really being a priority funding agency for the government.

## The Aftermath: Security Reimagined

September 11th changed not only my life but the entire focus of national security policy in the United States. For two to three years following the attacks, there was an unusual period where the national security community concentrated intensely on domestic security and the various challenges of building a more secure and resilient American society. In the maritime security domain, this meant significant focus on port security, transport security, and supply chain protection.

This period led to the formation of the Department of Homeland Security, a hodgepodge of organizations never coherently integrated. For the Coast Guard, which was beginning to acquire capability under Deepwater, being thrown into this departmental restructuring complicated the already challenging acquisition process. As the second-largest element of DHS, the Coast Guard found itself competing for resources against the Transportation Security Administration, which received substantial funding for aviation security, and Customs and Border Protection, which competed in similar operational spaces.

## The Coast Guard's Evolution in a New Security Environment

After working on the Deepwater program initially, I returned to an in-depth examination of the Coast Guard in 2011 and 2012, working with Admiral Gilbert, one of the founders of the Deepwater approach and one of the most respected senior Coast Guard officers ever to serve. We visited numerous Coast Guard facilities and wrote extensively about the organization's situation. The problems we documented then remain endemic to the Coast Guard: critical missions performed by dedicated professionals, with challenges consistently exceeding available resources, and perpetual uncertainty about funding priorities.

As administrations change and presidents come and go, the Coast Guard is repeatedly redirected to emphasize different parts of its large mission set, but without commensurate increases in funding. Each administration brings new style, emphasis, language, and focus, directing the Coast Guard to do various things differently while the fundamental resource constraints remain unchanged. The primary period I cover in this book, roughly from 2002 to 2016, represented a significant transition from the old Coast Guard to a Deepwater-enabled force. Even though Deepwater itself would disappear as a program, the acquisition wave it generated fundamentally reshaped the organization.

I bring the story up to date by comparing where we are now compared with where we were then. Unfortunately, not much has really changed in terms of resourcing and calibrating the mission sets facing the USCG.

## The Away Game Challenge

One of the critical themes that emerged from studying this period is the Coast Guard's role as an "away game" force. While the name suggests a focus on the coast and inner waterways of the United States Coast Guard has significant engagements abroad in the Pacific, Atlantic, and Middle East. This international dimension is frequently overlooked but strategically crucial.

In our work on rebuilding U.S. military capability in the Pacific, we started essentially with a focus on the Pacific Coast Guard and its particularly important role. As China has emphasized gray zone conflicts, operations below the threshold of conventional warfare, the U.S. Coast Guard, with its national security cutters and unique capabilities, represents a natural fit for addressing these challenges. The "white fleet" committed to maritime security and management, rather than war-fighting, offers distinct advantages in these ambiguous operational environments.

However, even as the Obama administration talked a great deal about the pivot to the Pacific, there was no increase in USCG funding to actually build a Pacific fleet of national security cutters, which were the platform of choice for such a role. This pattern exemplifies a recurring problem: strategic concepts and rhetorical commitments rarely translate into the resource investments necessary to make them operational realities.

The Coast Guard's away game functionality has increased in significance as we shift from what I call crisis management to chaos management. Traditional crisis management assumes discrete events that can be resolved, restoring a previous state of equilibrium. Chaos management acknowledges persistent complexity and uncertainty, requiring organizations to build adaptive capacity rather than seeking to restore stability. The Coast Guard's skill sets, its operational flexibility, law enforcement authority, and diplomatic character, are particularly well-suited to this evolving environment.

## The Pentagon Integration Challenge

An important but often neglected issue is the Coast Guard's relationship with the Department of Defense. The Coast Guard is also a DOD agency, yet the DHS move has made it more difficult for the Pentagon to grasp what the Coast Guard's role is in the national defense world. The 2025 shift to calling the Pentagon the "Department of War" certainly doesn't help from this perspective, as it emphasizes kinetic military operations over the broader security functions the Coast Guard provides. Deterrence and engagement in managing "gray zone" engagements are crucial and the Department of War label does not readily highlight this reality.

The Coast Guard serves as a key agency in dealing with the multitude of away game challenges. These include drug interdiction, illegal fishing, human trafficking, and maritime border security, essentially projections of globalization's darker aspects into the United States. The Coast Guard's unique position as a law enforcement agency with military capabilities makes it indispensable for deterring, defending against, and managing these transnational threats.

## Persistent Challenges in an Uncertain World

The fundamental challenges facing the Coast Guard have remained remarkably consistent over the decades. Personnel shortages persist, with recruitment and retention difficulties stemming from competition with private sector opportunities and quality of life concerns. Platforms and equipment suffer from chronic underfunding, leading to aging assets operating beyond their designed service lives. Infrastructure investment falls consistently short of what's needed, with Coast Guard stations, piers, and support facilities deteriorating while competing priorities consume available funds.

Perhaps most problematic are the constant swings in mission focus, which make it extremely difficult to fund a balanced Coast Guard.

Each administration arrives with its own priorities, whether emphasizing border security, drug interdiction, search and rescue, environmental protection, or national defense, and seeks to reorient the Coast Guard accordingly. Yet the underlying mission set doesn't shrink; responsibilities simply accumulate while resources remain constrained.

These endemic problems are not primarily about poor leadership or organizational dysfunction. Rather, they reflect fundamental tensions in how we conceive of national security and allocate resources accordingly. The Coast Guard operates at the intersection of multiple policy domains, defense, law enforcement, environmental protection, economic security, yet doesn't fit neatly into traditional bureaucratic categories or budget processes.

## From Crisis to Chaos: A New Strategic Framework

My professional focus after September 11th shifted from post-Soviet nuclear matters to engage with broader questions of terrorism, homeland security, and the fundamental reshaping of American defense priorities. The Cold War concerns that had dominated my career suddenly seemed like artifacts from another era, even though the nuclear dangers hadn't disappeared. They had simply been joined by threats we failed to adequately anticipate.

The defense community scrambled to reorient itself from state-based threats and conventional military competition to the challenges of asymmetric warfare, non-state actors, and the vulnerability of open societies to catastrophic terrorism.

- How do you defend against adversaries who don't field armies or navies but can turn civilian aircraft into missiles?
- How do you maintain the openness and freedoms that define democratic society while protecting against enemies who exploit precisely those freedoms?

These questions consumed the defense analysis community in the months and years after September 11th, representing a fundamental reorientation of strategic thinking. What has emerged is recognition that we face not discrete crises to be managed but persistent chaos requiring continuous adaptation. This environment, characterized by overlapping challenges, ambiguous threats, rapid technological change, and blurred boundaries between domestic and international security, demands organizational capabilities that traditional military structures struggle to provide.

The Coast Guard, with its operational flexibility, broad authorities, and institutional culture of adaptation, is better positioned for this environment than many larger and more heavily resourced defense organizations. Yet realizing this potential requires both adequate resources and clear strategic direction, precisely what has been lacking.

## Conclusion: Memory, Service, and Ongoing Obligation

The experience of September 11th is more powerful than any response to terrorism could be. It's the visceral disruption of ordinary life, the theft of security and normalcy. For those of us who were there, who felt the Pentagon rock, who saw the plane embedded in the wall, who smelled the burning for days afterward, September 11th isn't history. It's lived experience, as immediate and present as that pungent smell that still lingers in memory.

The slice of Coast Guard history I have documented in this book, from 2002 to 2016 and beyond, tells fundamental truths about the challenges this extraordinary organization faces. My book is dedicated to Admiral Gilbert, by way of remembering him and thanking him for the time and wisdom he gave me to understand the organization he served so admirably.

The Coast Guard continues to face personnel shortages, platform underfunding, infrastructure deficits, and mission focus swings that make balanced capability development nearly impossible. These are not new problems, nor will they be easily solved. They stem from the Coast Guard's unique position in our national security architecture, essential yet perpetually under-resourced, versatile yet constantly redirected, critical yet rarely prioritized.

As we navigate an era of strategic uncertainty characterized by gray zone conflicts, transnational threats, and the blurring of domestic and international security challenges, the Coast Guard's role becomes ever more important. The organization's ability to operate across the spectrum from law enforcement to military operations, its presence both domestically and internationally, and its unique character as a "white fleet" rather than a "gray fleet" give it capabilities that conventional military forces cannot replicate.

Yet capability without resources remains potential unfulfilled. The challenge for policymakers is not simply recognizing the Coast Guard's value but translating that recognition into sustained investment and coherent strategic direction. The alternative is continuing to ask more of an already overstretched organization while providing less than it needs, a trajectory that cannot be sustained indefinitely.

The smoke has long since cleared from Arlington. The Pentagon was repaired with remarkable speed. Life returned to something that looks like normal. But for those of us who lived through September 11th, the experience remains immediate, informing how we think about security, resilience, and the institutions we depend on to protect us. The Coast Guard is one of those institutions, imperfect, under-resourced, constantly adapting, yet essential to our security in ways that transcend its modest budget and public profile.

Understanding this organization's history, challenges, and potential is not merely an academic exercise. It is part of coming to grips with how we conceive of security in an age where threats are diverse, boundaries are permeable, and the distinction between war and peace has become increasingly unclear. The Coast Guard's story is, in many ways, the story of American security in the twenty-first century, ambitious in scope, constrained in resources, adapting continuously, essential always.

Even though there a many interviews which I conducted with Admiral Gilbert or by myself, there are many more which I had to not include in the book. A apologize to those whom I interviewed but did not include.

And in book published in 2026, I explore in more detail the nature of chaos management and its challenges, or the broader context within which the USCG operates.

A final note. With regard to the interviews and articles from the period when we conducted our interviews, I have indicated the date of the publication of the interview at the end of the article.

# PART 1

## COMMANDANT AND AREA COMMANDERS PERSPECTIVES

Part 1 explores the perspectives and strategic dilemmas faced by the Coast Guard's top leaders, specifically the Commandants and Area Commanders, during a period of profound transition and mounting operational complexity.

This part of the book provides first-person and interview-based accounts from Admiral Thad Allen (2010), Admiral Zukunft (2016), the Atlantic Area Commander (2011), and the Pacific Area Commander (2011), situating their insights within the Coast Guard's broader journey from the aftermath of September 11th through the challenges of the Deepwater modernization era and beyond.

1

# COMMANDANT THAD ALLEN, 2010

THE UNITED STATES Coast Guard stands at a critical juncture. With aging assets, a slowdown in procurement, and persistent budget shortfalls, the Service's ability to respond effectively to its expanding safety and security missions faces unprecedented challenges. The gap between operational demands and available resources continues to widen, creating a situation where the Coast Guard's legendary commitment to service increasingly outpaces its material capability to deliver on that commitment.

The ethos of the Service is such that its leadership would hesitate to acknowledge these limitations publicly. The Coast Guard has long operated under the principle of improvisation and dedication, finding ways to accomplish missions despite resource constraints. However, the reality facing the Service today is stark: the mismatch between assets and operational needs is growing, not decreasing. This trend, if left unaddressed, threatens not only the Coast Guard's operational effectiveness but also the safety of its personnel and the security of the nation's maritime domain.

The challenges extend beyond aging platforms. The Coast Guard operates in an increasingly complex environment where traditional maritime safety missions intersect with homeland security, counter-narcotics operations, and Arctic challenges. Each mission area demands modern assets with advanced C4ISR capabilities. Without sustained investment, the Service risks being unable to fulfill its statutory responsibilities.

## The Deepwater Modernization: Successes Obscured by Controversy

The very effective tools that the Deepwater and other acquisition programs have generated are not widely understood by policymakers or the public. The program, despite well-publicized management challenges and cost overruns in its early phases, has delivered critical capabilities that have already proven their worth in actual operations. The dramatic rescue off Cape Hatteras stands as a powerful example of what modern systems can achieve. That rescue would not have been possible without the C4ISR modernization implemented on the Coast Guard's HC-130J aircraft. The integrated sensor systems, advanced communications equipment, and real-time data-sharing capabilities enabled the aircrew to locate survivors in extremely challenging conditions and coordinate a complex rescue operation involving multiple assets.

Yet despite these demonstrated successes, the Department of Homeland Security has signaled its intent to slow down the acquisition of new maritime patrol aircraft. These platforms are essential for surveillance, security, and safety missions across the Coast Guard's vast area of responsibility. The HC-130J, in particular, represents a quantum leap in capability over the legacy HC-130H aircraft it replaces.

The newer aircraft features modern avionics, extended range, better fuel efficiency, enhanced sensor integration, and improved crew situational awareness. Delaying or reducing the procurement of these aircraft directly undermines the Coast Guard's ability to maintain persistent surveillance over critical maritime approaches, conduct long-range search and rescue operations, and support counter-narcotics interdiction efforts.

The situation is further complicated by competing demands for aviation assets. Military operations in Afghanistan and Iraq have strained Department of Defense airlift capabilities. Coast Guard maritime patrol aircraft and HC-130J transports provide valuable surge capacity for humanitarian assistance. The Haiti earthquake of January 2010 demonstrated this reality, with Coast Guard aircraft among the first responders providing critical lift and damage assessment capabilities.

## Haiti 2010: The Coast Guard as First Responder

On January 13, 2010, outgoing Commandant Admiral Thad Allen provided an overview of current Coast Guard challenges at the Surface Navy Association meeting. Much of his address focused on the Service's response to the devastating earthquake that had struck Haiti just hours earlier. Admiral Allen had been up throughout the night working with the Secretary of Homeland Security and other senior officials to coordinate the initial response to what would become one of the most significant humanitarian disasters in the Western Hemisphere.

The Coast Guard's response illustrated both the Service's capabilities and its operational philosophy. A Coast Guard cutter was first on scene in Port-au-Prince, with two additional cutters in the area quickly following. Another cutter deployed to Guantanamo Bay to await cargo shipments for transport to Haiti. This rapid response was made possible by the Coast Guard's routine presence in Caribbean waters conducting anti-drug, law enforcement, and migrant interdiction operations. These ongoing missions, far from being a distraction from disaster response, actually enabled it. The Coast Guard maintained capable assets in the region precisely because it executes multiple missions simultaneously.

The Haiti response also highlighted the critical importance of Coast Guard-Navy interoperability. Title 10 requirements mandate that the Coast Guard transfer to the operational control of the Secretary of the Navy when so ordered. This legal framework necessitates that Coast Guard and Navy systems, procedures, and personnel be capable of seamless integration. In Haiti, Coast Guard helicopters, HC-130 aircraft, and other assets flowed into the operational area quickly, working alongside Navy and joint force elements in a coordinated relief effort. The years of joint training, common communication systems, and shared operational procedures paid immediate dividends in this crisis environment.

Admiral Allen emphasized that the Coast Guard served as the first responder in Haiti, developing initial situation awareness and providing early assistance while larger military assets mobilized. The Navy and other services would necessarily follow with heavier lift capability and greater staying power, but the Coast Guard's ability to arrive quickly and begin operations immediately proved invaluable. This first responder role extends beyond disaster relief to encompass the Coast Guard's broader maritime security mission. The Service provides the persistent presence, local knowledge, and immediate response capability that larger military forces cannot replicate.

The Commandant also noted that the Service's execution of routine anti-drug, law enforcement, and migrant interdiction missions ensured the necessary presence of capable assets in the region. This observation underscores a fundamental principle of Coast Guard operations: the Service's various missions are mutually supporting rather than competitive. Assets deployed for one mission can rapidly transition to others as circumstances require. However, this flexibility depends entirely on having sufficient assets available. A Coast Guard stretched too thin to maintain adequate presence cannot surge effectively when crises occur.

Admiral Allen candidly acknowledged that while the Coast Guard had demonstrated its capability to surge quickly in response to the Haiti disaster, the Service was stretched to the limit. This statement, though understated, revealed the underlying tension between mission demands and available resources. The Coast Guard possesses well-developed response plans for mass migration events, plans that have evolved and improved since the Mariel boatlift of 1980 and have been regularly exercised. These plans involve extensive interagency coordination with state and local governments and organizations such as FEMA and the Red Cross. However, even the best plans require adequate resources for execution.

## Global Engagement and International Partnerships

Beyond immediate crisis response, Admiral Allen highlighted the Coast Guard's extensive international engagement. Most foreign navies function more like coast guards, focusing on constabulary missions, maritime safety, and border security rather than high-intensity naval warfare. This makes the Coast Guard an ideal partner for capacity building and security cooperation.

The Coast Guard supports Department of State initiatives through training programs, ship rider agreements, and joint operations. In counter-narcotics operations, ship rider programs allow foreign law enforcement personnel to embark on cutters, enabling legal interdiction in international waters. Admiral Allen specifically discussed anti-piracy efforts, noting that effective counter-piracy requires both stopping incidents and punishing perpetrators through proper evidence preservation and prosecution mechanisms.

## Recapitalization: A Strategic Imperative

Admiral Allen stressed that despite budget pressures, recapitalization must continue. He expressed concern about creating a hollow force where the Coast Guard maintains organizational structure while lacking platforms needed to execute missions effectively. This scenario would leave the Service unable to fulfill statutory responsibilities, increase risk to personnel operating aging equipment, reduce effectiveness in search and rescue, and diminish the nation's ability to protect maritime interests.

## The Arctic Challenge: A Case Study in Strategic Risk

The Commandant discussed the Arctic mission as a specific example of where the gap between requirements and resources poses strategic risks. Growing joint service and international cooperative emphasis on Arctic operations reflects the region's increasing strategic importance. Climate change is opening new shipping routes, enabling expanded resource extraction, and creating new maritime traffic patterns in waters that were previously ice-covered year-round. These developments generate new requirements for safety, security, environmental protection, and resource management.

Except for submarine operations, virtually all Arctic missions fall within Coast Guard rather than Department of Defense jurisdiction. These include exclusive economic zone enforcement, continental shelf boundary determination and protection, aids to navigation, pollution prevention and response, search and rescue, fisheries enforcement, and vessel safety inspections. The North Pacific and North Atlantic Coast Guard forums have provided useful models for international cooperation on safety, security, resource protection, and Arctic governance issues.

However, the Coast Guard's Arctic capabilities are severely limited. The Service operates only three polar-capable icebreakers: USCGC Polar Star, USCGC Polar Sea, and USCGC Healy. Polar Star and Polar Sea are decades beyond their intended service lives, with Polar Sea effectively non-operational. Healy, while newer, is primarily a research vessel. This small, aging fleet is inadequate for expanding U.S. interests in polar regions, particularly as Russia operates over forty icebreakers including nuclear-powered vessels capable of year-round Arctic operations.

## Conclusion: The Imperative
## for Sustained Investment

The Coast Guard activity discussed by Commandant Allen underscores the fundamental importance of presence in regional areas of operation, capable assets positioned in those areas, surge capacity to deal with crises, and robust C4ISR systems to enable coordination across the Coast Guard, with the Navy, and with partner nations.

There is simply no substitute for having capable assets on scene with detailed knowledge of local areas and highly trained crews. This capability requires a force in being, not a force in theory. The Coast Guard must maintain effective presence in its areas of responsibility with appropriate equipment and full interoperability to leverage what the nation and international partners can bring to core safety and security missions.

Yet the positive contributions the Coast Guard makes to national security, public safety, and international stability are not fully resourced. The Arctic mission provides perhaps the starkest example. With only three polar-capable icebreakers, two of them in dire need of replacement, the Coast Guard cannot adequately fulfill its Arctic responsibilities.

This capability gap exists across the Service's entire mission portfolio. Aging cutters, insufficient aviation assets, delayed modernization programs, and inadequate shore infrastructure all contribute to a growing divergence between what the nation asks of its Coast Guard and what that Service can realistically deliver.

The choice facing policymakers is clear. Sustained investment in Coast Guard recapitalization, continued modernization programs, and adequate operational funding will preserve the Service's ability to execute its diverse missions and respond to emerging challenges.

Alternatively, continued underfunding will create a hollow force, leaving the nation with a Coast Guard unable to fulfill its statutory responsibilities despite the dedication and professionalism of its personnel.

The gap between missions and resources will not close itself. It will only grow wider until deliberate action addresses the fundamental mismatch between what we ask of the Coast Guard and what we provide it to accomplish those tasks.

January 18, 2010

# 2

# COMMANDANT ADMIRAL ZUKUNFT, 2016

THE LAST TIME I met with Admiral Zukunft was when he was working at the USCG headquarters in California and responsible for USCG Pacific operations. Later he would become Commander, Coast Guard Pacific Area as well.

We had a chance to discuss his perspective on the way ahead for the USCG in his office in Washington DC on November 30, 2016.

## A Strategic Inflection Point

Obviously, the election of a new President and the formation of a new Administration will provide a new context for the USCG and its evolving role, but the emphasis which President-elect Trump has put on border security and defense and the decision to appoint General Kelly as head of the Department of Homeland Security could well provide an expanded context for the USCG.

The USCG plays a crucial role in Western Hemisphere security and defense and certainly General Kelly saw on a regular basis the crucial role the USCG has played in the region and shaping extended security and defense for US borders. Kelly's tenure as

Commander of U.S. Southern Command gave him firsthand experience working alongside Coast Guard assets in counter-narcotics operations and humanitarian missions throughout Central and South America.

For Admiral Zukunft, the USCG's role in Western Hemisphere security and defense has been significant, specifically as other DOD and security assets have deployed to the Middle East and the Pacific to deal with other global issues. This geographic displacement has created both challenges and opportunities for the Coast Guard, as the service has increasingly become the primary federal presence in regions that might otherwise see reduced American engagement.

## Intelligence-Driven Operations: A New Paradigm

The USCG today is quite different from 20 years ago as it operates now within an intelligence and operational web which provides a very different approach to deploying assets up against threats. The transformation from a primarily reactive service to one that operates with sophisticated intelligence integration represents one of the most significant evolutions in Coast Guard history.

According to Admiral Zukunft: *A key requirement for mission success is leveraging intelligence. We work intelligence across our agencies and internationally. This is crucial to provide risk informed decision-making.*

*A Coast Guard Cutter Stratton boarding team member inspects the bridge of a self-propelled semi-submersible interdicted in international waters off the coast of Central America, July 19, 2015. The Stratton's crew recovered more than six tons of cocaine from the 40-foot vessel. (Coast Guard photo courtesy of Petty Officer 2nd Class LaNola Stone)*

*We have constrained resources and we need to prioritize threats across the spectrum of operations. For example, in dealing with the afloat drug traffic we focus on the transit zones and work our ability to find choke points on the water and ashore to deal with the drug threat.*

*We work with a number of key governments in the Western Hemisphere to shape more effective intervention. If you look at DOD's statement of key priorities, they are not focused on the Western Hemisphere. We have the responsibility by default and design.*

*And a key part of homeland security is the security of the conveyer belt of maritime trade, which translates to about $4.5 trillion per year, which flow through our waterways and ports.*

This intelligence-driven approach has allowed the Coast Guard to maximize its effectiveness despite resource constraints. By focusing on transit zones and identifying choke points, the service can position assets where they will have the greatest impact rather than attempting blanket coverage of vast maritime areas.

## The Resource Challenge
## and Strategic Choices

Question: Clearly the limitations on resources is a key challenge but your approach allows you to get maximum return on investment by targeting the resources. How would you highlight that challenge?

Admiral Zukunft: *It is a challenge. The Navy's Perry-class frigates have gone away. On the best of days you have three Coast Guard ships in the Caribbean. That is your entire force to deal with threats in that region.*

*We have 80% awareness on the best of days, and perhaps we can target 10% of that drug flow.*

The stark mathematics of maritime security are sobering. With only three cutters available for Caribbean operations on optimal days, and the ability to interdict only a fraction of known drug trafficking, the Coast Guard faces an ongoing challenge of strategic prioritization. The retirement of the Navy's Perry-class frigates, which once provided additional presence in these waters, has further concentrated responsibility on Coast Guard assets.

## Recapitalization and the National Security Cutter Program

**Question: A key asset for recapitalization is your new offshore patrol boats. Could you discuss their role?**

Admiral Zukunft: *We have been struggling to get a program of record of the national security cutter across the finish line, and this is really the biggest acquisition for our service to provide the presence and enforcement assets which can provide for enhanced safety in security in our operations worldwide, but notably for extended border security for the United States.*

The National Security Cutter represents the Coast Guard's most significant modernization effort, designed to replace the aging Hamilton-class high endurance cutters. These vessels provide the operational range, endurance, and capabilities necessary for extended operations far from U.S. shores, making them essential for the Coast Guard's expanded strategic role.

## Legal Authorities: The Coast Guard's Unique Position

**Question: The USCG has a much greater role in security than generally recognized because of your legal authorities. Could you comment on this aspect of the USCG role?**

Admiral Zukunft: *We operate not just on our own vessels but we have a presence on the vessels of the US Navy and other nations. We are the only entity that has authorities to do anything about the security threats which we are prosecuting at sea.*

*The Coast Guard Cutter Forward returns to homeport in Portsmouth, Va., Monday, Jan. 7, 2013, after a 45-day patrol in the Caribbean Sea in support of Operation Martillo. The law enforcement crews aboard the Forward teamed with crew members from the Coast Guard Cutter Confidence and an embarked aviation detachment from Helicopter Interdiction Tactical Squadron to counter two illicit tracking interdictions leading to the seizure of more than $50 million in narcotics. (U.S. Coast Guard photo by Petty Officer 2nd Class Walter Shinn)*

*And we deal with a number of foreign governments in the Western Hemisphere who view the USCG as a key partner in shaping more effective Hemisphere security as well.*

*When you lay a map of the world flat and you look at where the USCG has authority, it reaches right up to the territorial seas in the countries surrounding the US and, in many cases, inside the territorial seas.*

*With the agreements we have now worldwide we do not have to wait till an anomaly in the manifest of cargo ship alerts us to a threat and simply have to wait till it shows up. We can intercept at sea and do a security check.*

This unique combination of military capabilities and law enforcement authority makes the Coast Guard an invaluable asset in situations where the use of conventional military forces would be inappropriate or create diplomatic complications. Coast Guard personnel routinely embark on Navy vessels and foreign ships precisely because they possess the legal authority to conduct law enforcement operations that naval personnel cannot.

## Arctic Operations and Future Challenges

**Question: With regard to the Arctic, there is an obvious need to ramp up US presence and the resources to provide for presence. How do you view the way ahead?**

Admiral Zukunft: *We clearly need a new icebreaker. We've written the operational requirement documents that make the icebreaker a floating command and control platform. We can put a skiff on it. It's also an instrument to enforce sovereignty.*

*Rather than ice hardened, you have actually an ice breaking capability up there as well. It is extremely hard to predict what that area's going to look like in 20, 30 years but without a new icebreaker we will be observers more than participants in shaping Arctic safety and security.*

*An independent High Latitude analysis confirmed that we need three and three – three heavy and three medium icebreakers.*

*We have helped stand up an Arctic Coast Guard Forum based on the Pacific Coast Guard Forum model. This allows the key national stakeholders in Arctic safety and security to work together where possible to enhance safety and security in this dynamic region.*

*We are looking to do a mass rescue exercise in 2017 around Iceland that will bring in Denmark and other NATO partners for a collective security effort. And to be clear, the USCG is the key sea service for the Arctic, the USN has in effect devolved Arctic security responsibilities to the USCG.*

The Arctic represents both a challenge and an opportunity for the Coast Guard. As climate change opens new shipping routes and increases human activity in the region, the Coast Guard's role becomes increasingly important. The establishment of the Arctic Coast Guard Forum demonstrates the service's commitment to multilateral cooperation in addressing the unique challenges of high-latitude operations.

## An Offensive Border Security Strategy

**Question: It seems your focus on borders is on a broader rather than narrower concept?**

Admiral Zukunft: *It is. Rather than having a goal line defense concept, we have an offensive strategy. When I think of a border, it begins at the territorial seas of the Pacific and Caribbean nations, which we deal with.*

*We have the ability to detect anomalies, we have authorities, and then when it comes down to the resources to be able to target that threat and meet it on an open playing field, rather than a goal line defense.*

*You might call this a layered defense strategy, but I prefer to call it an offensive approach whereby the USCG can leverage its authorities as far removed from the goal line as possible and practicable.*

*For example, I have discussed with the CNO the concept that we would create a permanent USCG presence in the South China Sea and related areas. This would allow us to expand our working relationship with Vietnam, the Philippines and Japan. We can spearhead work with allies on freedom of navigation exercises as well.*

This offensive approach represents a fundamental shift in thinking about border security. Rather than waiting for threats to reach U.S. shores, the Coast Guard pushes its operational boundaries outward, interdicting threats at their source or in transit. The proposed permanent presence in the South China Sea would extend this concept to the Indo-Pacific region, where the Coast Guard's law enforcement authorities and less militaristic profile could prove particularly valuable in addressing maritime security challenges while maintaining positive relationships with regional partners.

## Sustainability and Future Technologies

**Question: Clearly, you need more resources to expand presence, but the sustainability piece is often lost sight of. What are your thoughts on the sustainment piece?**

Admiral Zukunft: *There is usually much less focus upon sustainability but there are serious shortfalls which need to be addressed. We see our role as providing a key contribution to national security in dealing with non-state actors, whether is the threat of piracy, transnational crime or drug dealing.*

*The USCG provides unique authorities with titles 10 and 14 to provide for a unique instrument of security, particularly when one is looking at a more offensive approach to protecting our borders.*

**Question: Clearly, more platforms accompanied with better ISR and C2 is a key requirement moving forward, but what about the potential for robotic vehicles?**

Admiral Zukunft: *It would make sense for UUVs to be part of the USCG future, and we would start with the Arctic as a key area for such operations, to gain enhanced situational aware-ness in the region.*

The integration of unmanned underwater vehicles and other autonomous systems represents the next frontier for Coast Guard operations. These technologies could dramatically expand the service's ability to maintain persistent surveillance over vast ocean areas, particularly in challenging environments like the Arctic where human presence is limited by harsh conditions and high operational costs.

## Conclusion: The Coast Guard's Evolving Strategic Role

Admiral Zukunft's vision for the Coast Guard reflects a service in the midst of significant transformation. From its intelligence-driven operations to its offensive approach to border security, the Coast Guard is evolving to meet twenty-first century challenges with a unique combination of military capability and law enforcement authority.

The challenges are substantial: aging platforms requiring recapitalization, resource constraints that limit operational reach, and expanding responsibilities in regions from the Caribbean to the Arctic to the South China Sea. Yet the Coast Guard's flexibility, its unique legal authorities, and its partnerships with foreign services position it as an increasingly valuable instrument of American security policy.

As the incoming administration considers its approach to border security and defense, Admiral Zukunft's interview provides important insights into how the Coast Guard sees its role and the resources it needs to fulfill expanding missions.

The service's ability to operate far from U.S. shores while maintaining law enforcement authorities makes it an irreplaceable asset for addressing the complex security challenges of an interconnected world.

December 16, 2016

# 3

# THE ATLANTIC AREA
# COMMANDER, 2011

IN AUGUST 2011, Vice Admiral Parker was interviewed in his office in Portsmouth, Virginia, to discuss the challenges facing the U.S. Coast Guard in the Atlantic Area.

Vice Admiral Robert C. Parker assumed duties as Commander, Coast Guard Atlantic Area (LANTAREA) in April 2010, where he serves as the operational commander for all U.S. Coast Guard missions within a geographic region ranging from the Rocky Mountains to the Arabian Gulf and spanning five Coast Guard Districts and 40 states. He concurrently serves as Commander, Defense Force East, and provides Coast Guard mission support to the Department of Defense (DOD) and Combatant Commanders.

Before assuming command of LANTAREA, he served as the U.S. Southern Command's Director of Security and Intelligence in Miami, Florida. As the first Coast Guard officer to serve as a Director in any DOD command, he directed U.S. military operations and intelligence efforts and coordinated interagency operations in the Caribbean and Central and South America.

**Comment: The USCG divides the world into the Pacific command and the Atlantic command. We have**

**recently spoken to your counterpart, Vice Admiral Manson Brown, and now we are talking with you. Generally, folks don't realize that the Coast Guard is not a glorified harbor patrol and actually operates worldwide.**

Vice Admiral Parker: *A lot of people, including some folks in our own government, think "Coast Guard" and the word "coast" really jumps out at them. So they expect us all to be right here standing on the beach or within sight of land. Many people are surprised, including some of us when we came into the service, to find out just how worldwide the organization is and what all of our maritime operations are in the real world.*

*If you look at the issue in terms of protecting the marine transportation system and the flow of commerce, the big pieces we do inside our ports, in our inland waterways, and along the coastal routes, that's a big deal. That's a lot of daily activity that sort of goes under the radar of 99 percent of the American public, but it's 95 percent of the stuff that we do.*

**Question: How do you and Vice Admiral Manson Brown divide the world, so to speak?**

Vice Admiral Parker: *What Manson doesn't have, I do. While some would choose a river as a boundary, that is a transportation system for us, so we split at the Rocky Mountains and watch over the waters on either side of Central and South America, and then I go around and we meet on the other side of Africa and the Northern Arabian Gulf. The way it's split up in DOD's Unified Command Plan, I work with five of the six geographic combatant commanders in terms of their areas of responsibility.*

*So I have closer involvement with our overseas operations, and then port and waterway operations are done by our districts and sectors. We oversee their roles and look at trends, risk analysis, and balancing missions and resources there.*

*We spend a lot more time on migration, drug trafficking, and counter-piracy operations in Atlantic Area. Right now we have ships off the coast of Africa and in the Northern Arabian Gulf. I've got six patrol boats and seven crews and a port security unit helping out my U.S. Navy brethren and the Fifth Fleet over there.*

*I spoke with two of the fleet commanders yesterday, one here in Norfolk and the other one over in Naples who is AFRICOM's naval component and the Sixth Fleet commander. So I deal with these guys pretty regularly. It involves a fair amount of my time to work with that one partner (DoD). Because they're in places that we care about and most of our missions are global, their partnership is very important.*

*We have cryptologists who are over in Afghanistan and some other interesting places as well, supporting another mission. And then we have liaison officers all over the world. At any given time we're in about 42 different countries with either training operations or liaisons in this part of the world.*

**Question: What would you say to our readers who ask, "Why is our Coast Guard doing things all over the world?"**

Vice Admiral Parker: *In a nutshell, I would say we protect people from the oceans and the oceans from people, and this worldwide reach is necessary to protect the United States from threats delivered by the sea.*

*The oceans are worldwide, and they enable our economic system, so we can't view the oceans as moats that protect or delay threats anymore. The oceans also present vulnerabilities to our economy and our nation.*

*Let's look at one international organization as an example.*

*The International Maritime Organization (IMO), based in London, has about 170 states as members. They set standards for maritime safety, maritime security, and pollution prevention.*

*These three areas are vital to our success as a nation, and we must be active participants in all these programs. We must ensure that ships arriving here or sailing near our coasts are safe, secure, and environmentally responsible. Success in these areas requires informed participation with dozens of organizations and administrations.*

*Our vessels must comply with the same requirements as the safe, secure, and environmentally responsible vessels calling here. Absent this framework, international commerce would falter, and our economy would collapse with the consequential loss of millions of jobs.*

*Providing all this framework requires us to participate around the globe with knowledgeable people. Also, it requires fully responsive training delivered here and in host countries when they don't have the resources or experience to do it yet.*

*Additionally, most navies throughout the world are much more like our Coast Guard than our Navy. They do much of what we do, save lives, enforce laws, guard against illegal migration, protect fisheries, and execute environmental prevention and recovery programs.*

## Question: Could you explain the difference among the command levels?

Vice Admiral Parker: *The districts are more tactical and at the lower operational level of things. And clearly at the sector they are very tactical and very hands-on in mission execution. By the time you get up here, we're very much in the operational world, balancing risk across missions.*

*So most of what we do is plan for contingencies and crises and then figure out how to have the best effect across the various regions and missions with the resources we have. So I have a lot of oversight. But I have some operational things our staff works directly, as I discussed above.*

## Question: In a period of financial constraints and strategic redefinition, what does the USCG bring to the table?

*Vice Admiral Parker: When times get tough, people tend to retrench into a more defensive posture, which makes it tougher sometimes to get in the national collaborative mode. Such collaboration is in our DNA. This may also explain why we wind up in leadership positions in these large operations, when things go bad we look around to see who we have and who can be helpful and then get after it as a team.*

*Whether that's presence, authorities, capabilities, or capacity, whatever it is, if it works we want it. We're sort of agnostic about those things.*

*We have our own moments where we want to do it ourselves, but we just live in a world where we simply must collaborate, whether it's in the ports or offshore or whatever. And we are eager to tap into national systems. So we're naturally curious to go around and see who else can help there. The other thing that is often undervalued is the vast portfolio of authorities that we bring to the table.*

## Question: Could you discuss the challenges of maintenance and keeping the older fleet operational?

*Vice Admiral Parker: The sustainment challenge across the board is evident everywhere I go. When I travel, I see the same thing. I see old stuff, sometimes relieved to find new stuff, but still see the herculean challenge faced by the individuals trying to maintain those older platforms.*

*We need to do a service life extension program on the 140s. *The inland waterways assets need replacement, river tenders and construction tenders that mark our dynamic marine highways in the heartland. There's no relief on the horizon for those, and I just don't know how you squeeze that in with the priorities here right now and the fiscal realities.*

**Question: How do you think we need to shape an ability to get the different players on the same page in a crisis? How do we leverage IT and other systems to get the decision-making focus on the right page?**

**You really need a collaborative knowledge system to shape a proper C4ISR response. For us, the point of C4ISR is to enable better decision-making, not to just collect information for maritime domain awareness in and of itself.**

Vice Admiral Parker: *I was afraid you were going to fall in that regular trap that I have to face all the time. And I was afraid you were going to stop at the IT thing, because that's where everybody runs to: thinking that the IT is the solution, and it really isn't.*

*You don't want the IT guy to be the one that does your knowledge management. It's like giving over control of the architecture of your house to the plumber: you have all the toilets right inside the front door and it's locked. Completely unuseful to the person in the house, but very convenient for the plumber who wants easy access and wants to keep that part of the house secure.*

---

*   This is the Ice Breaking Tug Class vessel; there are nine vessels in the class; most are stationed in the Great Lakes.

*So in the knowledge/information/data world, the IT may actually work in day-to-day activity, but you know what happens in a crisis? We can't pull what we need in a way that is helpful under stress and compressed decision cycles, or it is so secure we can't get it in the hands of our first responders on scene.*

*You're going to make decisions anyway, and if they're ill-informed, or worse, they're informed by bad data, you're really in difficulty when these events unfold on flash media on a national or world stage.*

*The importance of getting the right information quickly has been illustrated by the recent Gulf oil spill. The national response center command center is supposed to keep track of all the boom we have in order to respond. There are a lot of different types of boom, as it turns out.*

*But we needed to know specifics such as: tell me where all my 18-inch hard ocean boom is, because that was the big thing in demand. That was the coin of the realm in the decision-making center. How many miles of this do we have?*

*You know, pick a number. Didn't matter, the number was wrong because it was an aggregate of going to every oil spill response organization in the country saying "how much boom do you have?" And they're obligated to have on-call for an operation "Y" amount of boom.*

*So they would go out and contract with contractor "X" and say, "I am obligated to get 10,000 feet of boom within 24 hours, can you meet that so I don't have to warehouse it?" And they would all say, "yes."*

*So seven different organizations all go to the same guy, so when we get done, "okay I need 200,000 feet of boom," and you get 10,000 because it was all the same guy counted 20 times.*

*OK. We have got a real problem here: how fast can you make this stuff?*

*But when you think back through the IT approach, it violated the first rule of information: data management. Where is the authoritative source for that data and how do you manage that?*

*This was similar to the same thing I had down in SOUTHCOM during Haiti. We just could not tell the chairman how that bottle of water got from Keokuk, Iowa all the way down to that Haitian's hand.*

*It seemed irrelevant to me, but he wanted to know this information. But I had no way to track that.*

*So we went to folks like FedEx, DHL, and UPS and said, "How do you guys do this?"*

*And that's where I learned about authoritative source. Authoritative source for every bit of data is hugely important. So how many authoritative sources do you think roughly we would have in a Coast Guard? So I asked this question. The answer I got: 362. There is absolutely nothing authoritative about that number.*

*And what we haven't done as operators is put an opera-tional requirement demand on information to make decisions in crisis. The core need is to aggregate information during crisis. This is a huge issue for us and is mind-numbingly hard to get under control, especially in times of tight budgets.*

**Question: Your focus is upon building proper knowledge management tools. Could you explain your thinking about the way ahead?**

Vice Admiral Parker: *To repeat, the knowledge manage-ment piece really gets confused oftentimes with the IT, the technology side of things. And that's just the tool.*

*You have to design with the focus upon having to make decisions during contingency and crises as your center point before you go into the design of all these systems. Everything we have, whether it's personnel or logistics, all these systems are built for day-to-day operations and activities, and they tend to fail us during contingency and crisis.*

*So the first thought you have to get into people's brains is: what are the decisions that I'm going to have to make during crisis, what contingencies, and what information do I need to have? And then how do I feed that information and catalogue that on a regular basis in a way that I can retrieve that easily and with great veracity and repeatability?*

*Which was a real problem for us during the Deepwater Horizon response, because if you have more than one source the data doesn't upgrade at the same pace; you've automatically got a problem in terms of whether people believe you or not.*

*I think the first thing we must do is change the mindset to get away from just letting every individual program run where their information goes and figure out what it is that serves the mission. The best way to test that I think is to stress-test it and put it in a contingency and crisis.*

*Whether we're doing it during exercises or lessons learned from actual events and operations, I think we really have to go back and look at what decisions we need to make and where that information resides and then how we collect that as a whole service.*

**Question: And there is obviously a cybersecurity part of this problem as well?**

Vice Admiral Parker: *The Marine Transportation System vulnerabilities are just unknown in the cyber world. So I'm teasing that out.*

*I'm doing a number of different speeches and a couple of different articles between now and November. I'm trying to excite the nerve in different places. I've gone around and I've talked to the Houston Port partners and I've talked to New York port partners at length, and what is clear is that everybody has become accomplished in thinking of security in terms of what's inside my fence line and my facility. Or, you know, ships, or the places they go. And then what the external physical threats are.*

*Cyber is very different, and it doesn't take somebody doing a malicious act to break up systems now. With the automatic control systems you have at container terminals or locks and dams, or any number of things, a loss of power, phone line, or computing capability can have a cascading effect through this system.*

*So we don't fully understand what their accidental interdependence is in the cyber world in ports. So just understanding that, exploring that a little bit to understand what the vulnerabilities are, never mind what threats might be, but just knowing where the vulnerabilities are I think will be helpful.*

*We're trying to pick a few ports that are interested in doing this. Houston certainly was interested, and it's a place where the port partners have already worked past the competitive thing to understand that if one falls, all fall. They have clearly figured this out. And it's very impressive to see.*

**Question: You recently co-authored a piece with Vice Admiral Manson Brown on the need for the National Security Cutter. Could you address how the NSC fits into the Atlantic area strategy? Or put a different way, if you received three NSCs tomorrow, what would you do with them?**

Vice Admiral Parker: *Well they'd be welcome and immediately helpful. Both because we have a shortage of ships because things are getting old, and we have serious sustainment issues for the current fleet. The youngest ones are over 20 years old, and that's not young.*

*When I give change of command speeches, to give perspective I talk about what life was like when these ships were built. And the cheapest gas from the newest platform when they were built was a dollar twelve, and the cost ranges down to 32 cents a gallon. Many of the ships were built before the legislation that impacts the bulk of their current mission.*

*No question our ships are all getting old. But beyond that, the National Security Cutters certainly represent more than just a replacement for high endurance cutters in the USCG. It can and will have capabilities when it's finally flushed out that will create a much larger impact and effect to coordinate operations, not just be a point in the ocean from which you operate, but from which we can control sizable pieces of the ocean on waters that we regulate.*

## Question: If you had them for the Haiti events, how would you have used them?

Vice Admiral Parker: *That would have been huge. The Coast Guard of course was first in due to our regular presence there, but we couldn't sustain it. Certainly DOD brought a lot of muscle to the operation.*

*If I'd had an NSC at the front end of Haiti, the USCG could have controlled the air space in there as well for the critical first days. They would have had a full appreciation for what was going on in and around the littorals, and you could have done a lot of command and control for all the little different parties we were putting on the beach because comms were gone.*

*They were just completely out down there. Best thing I had when that happened as the J3 down in SOUTHCOM was we had a cell phone with the aide to the deputy commander who was in the earthquake that lasted about 36 hours and then it went dark.*

*That was kind of an awkward period there where the ship down there was the best way I had information, but what they didn't have was connectivity back to different people.*

*With the NSC you will have a significant bubble around the ship for C4ISR, and with the bigger bubble you have a much greater ability to control things. And that is true whether that's further down range or in a place where you've stripped out communications and command and control architecture that normally exists. Whether it's a Katrina, or a Haiti, or any other calamity where you lose comms and related capabilities.*

*Another key piece is to understand the aviation and related assets which you can fly off of the ship, which gives you greater range and operational capability. A ship deployed forward without a helicopter or without a maritime patrol asset overhead really controls only a 12-mile radius of ocean. With the helo, you can get that out to about 75 miles pretty cleanly, and then you put the over-the-horizon boat in the mix, your actionable radius goes out to about 100 miles.*

*The speed and endurance of the NSC operating within a much larger C4ISR bubble allows the USCG to increase significantly its operational area and its ability to anchor the AOR, whether sea-oriented or land-oriented.*

Vice Admiral Parker During the SLD Interview
(Credit: SLD)

By Robbin Laird and Ed Gilbert
October 31, 2011

# 4

# THE PACIFIC AREA COMMANDER, 2011

VICE ADMIRAL MANSON BROWN, the Coast Guard Pacific Area Commander, assumed command of Coast Guard Pacific Area in May 2010, where he serves as the operational commander for all U.S. Coast Guard missions within half the world, from the Rocky Mountains to the waters off East Africa's coast. He concurrently serves as Commander, Defense Force West, providing Coast Guard mission support to the Department of Defense and Combatant Commanders.

The Admiral brings varied and distinguished experience to this role. His previous commands include the 14th Coast Guard District, Maintenance and Logistics Command Pacific, Sector Honolulu, and Group Charleston. From 1999 to 2002, he served as Military Assistant to the U.S. Secretary of Transportation, including six months as Acting Deputy Chief of Staff following the September 11, 2001 terrorist attacks. In 2003, he served as Chief of Officer Personnel Management at Coast Guard Personnel Command. From April to July 2004, he was temporarily assigned as Senior Advisor for Transportation to the Coalition Provisional Authority in Baghdad,

Iraq, where he worked in a combat zone overseeing restoration of Iraq's major transportation systems, including two major ports.

Earlier tours of duty include Assistant Engineering Officer aboard the icebreaker GLACIER, Project Engineer at Civil Engineering Unit Miami, Deputy Group Commander at Group Mayport, Florida, Engineering Assignment Officer in the Officer Personnel Division at Coast Guard Headquarters, Facilities Engineer at Support Center Alameda, and Assistant Chief, Civil Engineering Division at Maintenance and Logistics Command Pacific.

## The Pacific Theater: Scale and Economic Stakes

The discussion began with the sheer enormity of the Pacific and its importance to the American economy.

"Most people don't realize that 85 percent of the U.S. exclusive economic zones are in the Pacific, mostly in the Central and Western Pacific," Admiral Brown explained. "There are many economies in that region driven by the fishing industry."

He emphasized a critical reality of marine resource management: "Even with good enforcement in U.S. EEZs, the fish know no boundaries. They shift from our EEZs to those of other nations and can be overfished there." To address this challenge, the Coast Guard formed partnerships with adjoining countries working their EEZs to manage illegal fishing beyond U.S. waters.

"We developed a joint strategy, a ship rider program where we use Coast Guard assets and embark enforcement officials from six nations that have signed ship rider agreements," the Admiral noted.

The distances involved are staggering. Admiral Brown pointed out that the Central and Western Pacific is significantly distant from the continental United States, with sovereign American territory located throughout the region. "To deploy a Cutter from Alameda, California, our Pacific headquarters, to American Samoa requires ten or more days," he explained.

## The Endurance Imperative

This geographic reality drives a fundamental operational requirement: endurance. If it takes more than a week each way, endurance translates directly into operational time on station. The Admiral emphasized another crucial factor: "You don't have the infrastructure in the Pacific that you do in the Atlantic. In terms of pier space, fuel, engineering support, food and other logistics, you have to take it with you. When you're in a place like American Samoa, you better have most of what you need to operate."

Weather conditions add another layer of complexity. "As a former icebreaker sailor, I can tell you that Pacific storms can whack you pretty heavily," Admiral Brown noted. "We need substantial ships to protect our crews and promote mission efficiency. And when folks are in Hawaii, they forget that the seas are rough around the islands. This is not the Caribbean."

He provided a stark illustration: "If you go a mile and a half offshore to do a SAR case in Hawaii with a 25-foot rigid hull inflatable, you're doing a SAR case in the middle of the Pacific Ocean. People think of Hawaii like Florida with the protection of the land mass, but there is no protection out there. You will be operating in seas that will scare you."

## Economic and Security Implications

The economic stakes are enormous. "These areas are some of the richest tuna fisheries on the planet," Admiral Brown stated. "We'll see a collapse of our fisheries if we don't protect these regions, which will affect the fishing economies throughout adjacent areas. There are 22 small nations of Oceania whose economies are driven by fishing licenses and fish. If those fisheries collapse, we could potentially see Somalia-like instability conditions closer to our sovereign territory."

Physical presence is non-negotiable. As the discussion turned to the importance of cutters, the Admiral was emphatic: "If you physically are not there, you can have all the ISR you want, but that's not a good deterrent to anybody. And it doesn't allow you to prosecute."

"It's presence in a competitive sense," he continued. "Because if we are not there, someone else will be there, whether it's illegal fishers or Chinese influence in the region. We need to be very concerned about the balance of power in the neighborhood. If you take a look at some of the other players operating there, there is clearly an active power game going on."

The Admiral provided a concrete example: "When I was in Tonga, I observed large structures built by the Chinese government and am watching other nations expand their influence around the world." The message is clear: not being present sends its own signal.

Admiral Brown identified another dimension of national security concern: "If you take a look at the march of terrorism through places such as Indonesia, it's not too difficult to craft an instability scenario where it could leap to Oceania, allowing our enemy to potentially get closer to reach out and touch us. I remind people that even though American Samoa is a U.S. territory, once you get to American Samoa, you're in America. It's not too difficult to reach out and touch us from there."

## The North Pacific Coast Guard Forum

As the U.S. representative to the North Pacific Coast Guard Forum, Admiral Brown offered insights into this important multilateral mechanism. "As I reflect on the history of the North Pacific Coast Guard Forum, one should remember it was a U.S. Coast Guard influenced effort," he explained. "The forum allows us to establish the U.S. Coast Guard as the honest broker for protection of fisheries in the high seas drift net area of the Northern Pacific Ocean."

The collaboration attracts interest from China, Russia, Japan, Korea, and Canada, the forum's partners, enabling mutual account-ability for fishing fleets operating in the area and collaborative enforcement efforts. "If you extracted the presence of the U.S. and the U.S. Coast Guard, instability would result," Admiral Brown stated. "We are really the glue that holds that forum together."

Having attended three different forum meetings, the Admiral was unequivocal: "It is not an understatement that we play a crucial leadership role. This is due to the respect we garner from those other nations and the capability we bring to the table."

## Unique Capabilities and Multi-Mission Nature

Admiral Brown outlined three capabilities that separate the Coast Guard from the Department of Defense: regulatory capability, law enforcement capability, and emergency response capability. "That really gets to our multi-mission nature," he explained. "Even though we may be conducting combined operations with Canada, Korea, Japan, or China, we're also there just in case something goes wrong so we can intercept the problem and provide search and rescue capability."

The concept of forward deployment takes on special meaning in the Pacific. "With the vast distances, what we refer to as the tyranny of distance, if you don't have enduring presence, which cutters bring, then you're not going to get to where you may be needed in an emergency in time, particularly in a place like the Bering Sea," the Admiral noted.

Pre-positioning becomes essential. Operations in the Bering Sea enable a positioned emergency response asset for search and rescue or pollution response. "There is no significant logistics support up there to enable rapid deployment," Admiral Brown explained. "Air only gets you so far. You need an emergency surface asset to pull it all together."

# Layered Strategy and International Engagement

"Many people believe we need to be a coastal coast guard, focused on ports, waterways, and the coastal environment," Admiral Brown observed. "But the reality is that because our national interests extend well beyond our shore, whether it's our vessels, mariners, possessions, or territories, we need to have presence well beyond our shores to influence good outcomes."

As both Pacific Area Commander and Pacific Fleet Commander, Admiral Brown oversees both the close-in game and the away game. The away game includes tangible authorities and capabilities like fisheries enforcement and search and rescue presence, but also softer capabilities. "We do a lot of nation building. We perform a lot of theater security cooperation for PACOM. We'll send ships to Japan and China just to exchange ideas, discuss common objectives and capabilities, and demonstrate American engagement in the region."

International respect for the Coast Guard stems from multiple sources. "The USCG is respected internationally because of our law enforcement and regulatory capabilities and our history," the Admiral noted. "When people see our response to Katrina or Deepwater Horizon, they want a piece of us."

The unique character of the Coast Guard simultaneously a security and defense entity enables a broader dialogue than a pure military force could conduct. "It comes down to common interests," Admiral Brown explained. "The common interests are maritime safety, security, and stewardship. Other nations understand we're also a military service, and we play that security-defense interface, but that's not how the conversation starts. They're interested in protecting their shores, shipping, ports, waterways, and environment."

## The Power of the Uniform

The Admiral emphasized that the Coast Guard's military character remains crucial to its effectiveness. "Part of our framework of respect and credibility is the fact that we wear this uniform," he stated. "People are intrigued in the international community by us. Our unique military and law enforcement character, combined with this uniform, makes it work for us. If I had gone to Beijing in a suit, I would've had a very different reaction."

He drew on his 2004 experience in Iraq: "When I was working for Ambassador Bremer, it was a civilian position. But I took along my Coast Guard uniforms, and it didn't take too many days for me to figure out that I better wear the uniform because it's a symbol that commands respect within the international community. That's something that cannot be lost in the discussion about the future of the USCG and its role in the Pacific."

## The Arctic Challenge

The conversation concluded with the growing Arctic challenges. "The Arctic presents a series of predictable surprises to us," Admiral Brown warned. "One of the things that keeps me up at night is that one of the cruise ships, adventurer cruise ships with 1,000 people onboard, are going to dip their nose into the Bering Sea and have a mishap. We will not be present to craft a response, and we do not currently have the infrastructure to help those in distress."

He noted increasing international activity: "The Russians are engaged in the Arctic, and even the Chinese are building an icebreaker. There is more and more human activity in the Arctic because of more and more water being open." The fundamental challenge is stark: "Once the ice becomes water, we have the authority, but we don't have the capability. And we don't have the infrastructure to do the job. It's not just the Coast Guard's problem; it's a national problem."

The Admiral pointed to other nations' preparations: "If you take a look at what Canada is doing to build infrastructure in anticipation of an opening Arctic, we're at least ten years behind them. If you take a look at Russia and what they're doing, they at least have a plan."

Resource competition adds urgency. "When it gets to the law of the sea and the plotting of territory up there for resource exploitation, if we're not careful, we're going to have our backyard picked, and we won't be able to do anything about it because we won't have the investment or infrastructure to support Arctic engagement."

Admiral Brown outlined requirements for Arctic capability: "We need a mix of capability. We need to reinforce our icebreaker fleet now and then make sure we have a shipbuilding strategy that allows us to have ships with hardened hulls and other unique capabilities that allow them to operate in areas with ice."

Critically, operational skills cannot be allowed to atrophy: "It takes ten years because of the challenges of ice breaking to grow a good icebreaker sailor. If we let our current ice breaking fleet atrophy in this period of uncertainty, then we start at a risk position to try to grow that capability for the future. I am all for hedging our bets by investing in what we have just to keep our hand in the game."

## Strategic Imperatives

Vice Admiral Brown's perspectives illuminate the Coast Guard's indispensable role in Pacific strategy. The service's unique combination of regulatory, law enforcement, and emergency response capabilities, coupled with its military character and international credibility, positions it as an essential instrument of American power and influence across the world's largest ocean. From fisheries protection and maritime safety to great power competition and Arctic sovereignty, the Coast Guard operates at the intersection of economic security, environmental stewardship, and national defense, a role that will only grow more critical as Pacific challenges intensify in the years ahead.

Vice Admiral Manson Brown during the interview,

By Robbin Laird and Ed Gilbert

August 13, 2011

# 5

# ANOTHER STRATEGIC CHALLENGE: THE NORTH PACIFIC COAST GUARD FORUM

THE LITTLE KNOWN North Pacific Coast Guard Forum is a key platform from which the Pacific powers can shape collaboration to enhance maritime safety and security in the Pacific.

Established in 2000, the Forum brings together six nations, the United States, Russia, Japan, South Korea, Canada, and China, in a unique cooperative framework that transcends traditional military-to-military relationships. By operating through coast guard organizations rather than navies, the Forum has created channels for dialogue and cooperation that might otherwise be impossible, particularly with China and Russia.

The Forum represents a practical response to the growing complexity of maritime challenges in the North Pacific. From illegal fishing that threatens marine ecosystems to drug trafficking, search and rescue, and environmental protection, the nations bordering the Pacific face common threats that require coordinated responses.

In December 2010, we sat down with Rear Admiral Bob Day to discuss his experience with the Forum. Rear Admiral Day is now based at USCG Headquarters, but he had seven years of experience with the Forum from 2002 through 2009.

At the heart of the effort is to find ways to get the various players to network their capabilities to take on the maritime security and safety challenges in the vast Pacific.

## The Players: A Diverse Maritime Coalition

**Question: Before you discuss the Forum, could you give us a sense of the kind of Coast Guards that exist in the North Pacific? Who are the big players?**

Rear Admiral Day: *Well, obviously, we've had a long-term relationship with Japan. Japan's Coast Guard is formulated very much after ours and the Korean Coast Guard also bears similar resemblance to USCG organization and missions. These relationships date back decades, with both Japan and South Korea modeling their coast guard organizations on American principles following World War II and the Korean War respectively. The institutional similarity makes cooperation more natural, as we share common organizational structures, training approaches, and operational concepts.*

**Question: And both have substantial assets?**

Rear Admiral Day: *The Japanese have very significant vessel and air assets. South Korea also has a very robust coastal fleet and aircraft; the Korean assets tend to operate primarily within their EEZ. Both nations have invested heavily in maritime security.*

*The North Pacific CG Forum has conducted numerous exercises and several joint operations and each of the members have contributed assets or personnel. Japan and Korea, because of the coastal nature of their vessels, have generally provided air (Japan) or personnel (shipriders) for operations being conducted on the high seas.*

## Question: What about the other players?

Rear Admiral Day: *The Russians are significant players. They have a Coast Guard, but it is part of the Federal Border Guards. And only recently have they called themselves a Coast Guard. As a matter of fact, the Generals who lead both the Border Guards and the Coast Guard department are here in Washington [at the time of the interview] for discussions with the Commandant and other U.S. agencies.*

## Comment: And the Russian Border Guard certainly has a deepwater fleet.

Rear Admiral Day: *Absolutely, they have a lot of water in their EEZ to cover and they're out there patrolling all the time. As a matter of fact, they're probably the major Forum member that comes out in deepwater operations. The Russians generally provide a vessel that's a little larger than a 378 (a 378 foot Coast Guard High Endurance Cutter). The Chinese Fisheries Law Enforcement Council (FLEC) has also brought a vessel out for high seas combined operations during the past several years.*

## Question: And then you use their air assets as well?

Rear Admiral Day: *As a matter of fact, the Russians always do provide some air support. The Japanese have also been providing air support. They use a Gulfstream G5. And they've been flying that out for the North Pacific joint patrol for high sea drift netting.*

## Question: What about the South Koreans?

Rear Admiral Day: *The South Koreans have generally been riding aboard our 378's, primarily with a ship rider. They have not brought air assets out.*

**Question: But the Japanese do?**

Rear Admiral Day: *The Japanese have brought out their G5. But they have not brought vessels. Most of their vessels are very much coastal oriented. Most of them are jet-powered and go very fast, but they don't have a lot of endurance. So going out and spending a long time on the high seas is generally not a capability that they have a ton of. But boy, they can swarm all over their coast and cover their islands and everything like that fairly fast.*

**Question: So basically, it's the Russians and we who have a deeper sea capacity?**

Rear Admiral Day: *Well, the Chinese have got it as well with the FLEC vessels that have been participating. The Chinese don't necessarily have a coast guard. Their coast guard like organization exists under the ministry of public security. And when we meet, we're generally meeting with their personnel from the ministry of public security. Then there's another element in China, which is their fisheries law enforcement group, and so we meet with them as well too, they're generally part of it. And they have some vessels, and the ministry of public security has some vessels.*

**Question: And what about the Canadians?**

Rear Admiral Day: *Canadians are generally providing CP-140 Aurora aircraft. They haven't provided any patrol vessels, because their coast guard does not have a ton of long-range patrol. They have patrol boats, hovercrafts and buoy tenders.*

## Building Interoperability: The Foundation of Cooperation

**Comment: So the folks who participate in this Forum have different capabilities, different assets and probably different concepts of operations for their Coast Guards. So part of the task of such a forum is to figure out how to knit this capability together to deal with some common missions.**

Rear Admiral Day: *That's exactly right. Some of the early pieces that before we even started doing joint operations, there's several elements that you need.*

- *How are we going to do operations when we're together?*
- *What are our communication frequencies?*
- *What are the procedures we're going to use?*

*So we created what was called a combined operations group. This workgroup had representatives from each of the countries and formulated the processes and procedures so that we could effectively operate together for a wide variety of scenarios. I think we are on the second or third version right now of the combined operations manual. And it keeps getting refined and better each time.*

*The combined operations manual represents years of work to overcome significant differences in equipment, procedures, and operational culture. The manual provides a common framework that allows ships and aircraft from different nations to work together seamlessly during operations.*

*The other piece that they needed was we needed an information exchange system. How do we talk to each other?*

*And that was probably the piece when I got involved in early on is what kind of computer system are we going to come up with such that we can exchange information. And do to so with a look towards how do we operate with each other? The Russians developed and host a system called the North Pacific Coast Guard Automated System which uses a certificate based system to encrypt and have a common platform to talk on. And it's web-based.*

*And it's gotten to the point now, particularly, that D-17 (the Coast Guard's Alaska Command) and the Russians talk to each other every day on it. They exchange information on vessels that are in the Bering Sea. And so, it's moved up into that, it has chat capability, such if you want to get a bunch of people up together, you can chat back and forth basically using the chat function.*

*There's a database there that they've been building over time that has vessels of interest. And such that the Russians can populate it, anybody can populate it saying hey, we caught these guys; these guys were doing illegal fisheries, this is what we caught them doing, et cetera.*

*The information exchange system has proven to be one of the Forum's most significant achievements. In an era when information sharing between nations often faces bureaucratic obstacles, the Forum created a secure system that enables real-time operational coordination.*

*The daily exchanges between the Alaska Command and Russian counterparts demonstrate how the system has moved beyond formal meetings to become an operational tool.*

*Fisheries have been a major driver, but they're also committees that look at how can we do cooperative work on migration. How can we do cooperative work on counter narcotics? Again, they've got some significant major problems there too.*

*All of the North Pacific Coast Guard Forum partners face the challenge of trying to stop illicit drugs from coming in or precursor materials too. And so, trying to get inside of those and disrupt those rings as well is another major issue we're dealing with.*

## Structure and Evolution of the Forum

**Question: When did you first get involved with the forum and for how long?**

Rear Admiral Day: *2002 was my first year in the Forum and was involved until I left the Pacific in 2009. Seven years. I started out just being a member of a workgroup and then moved into the position where I was the Head of the U.S. delegation for numerous experts meetings.*

*There are two major meetings each year. The first meeting is what's called the expert's meeting. And that's where the workgroups really get together and generate materials for review later on, developing the combined operations manual, developing the information exchange system. The expert's meeting is generally conducted in the March/April timeframe.*

*And then, in September/October timeframe is what they call the summit. And that's when the Commandants or the Commandant equivalents for each one of these countries gets together, reviews the material that's been done by the workgroups, and then approves it or gives them additional direction on things to work on.*

*The Summit is where we sign cooperative agreements once we've got something finalized or approved like the combined operations manual.*

*That's the cycle.*

*The workgroups are working electronically, moving stuff back and forth in the meantime. And then they have their meeting, refine it, and then present it at the Summit. The Summit is really the culmination of it, and it's really quite a big event.*

*You've got to bring six nations together; you've got have translation services; you've got to have all the hosting and all the rest of the cost with it. It's a pretty heavy lift.*

**Question: Who pays for the meeting?**

Rear Admiral Day: *We have to pay for it out of hide. We get no extra money to host the Forum.*

## Looking North: The Arctic Challenge

**Question: And presumably the Arctic as a game changer can be managed in part by the existence of this Forum?**

Rear Admiral Day: *Having set this Forum up, it'd be a very useful toolset that to manage some of these new dynamics. We're constantly shifting the content inside the North Pacific Coast Guard Forum. And the Arctic is starting to become a major discussion point.*

- *How do you do search-and-rescue up there?*
- *How would we coordinate a Russian, Canadian and U.S. response to a major disaster?*
- *And what happens when you have a 3,000-person cruise ship that gets into trouble?*

*And they're miles and miles away from anybody who can respond. And it's going to take a response from all those nations that are around that perimeter probably to get there and deal with it. And so, starting to figure out those frameworks on how you're going to do that. How do you deal with a spill of major significance in the Arctic?*

*The Arctic is emerging as perhaps the most important new focus for the Forum. Climate change is opening previously ice-covered waters to shipping and resource exploration. A major incident in the Arctic would require cooperation among all nations with Arctic interests, making the Forum's existing relationships and procedures invaluable.*

## Strategic Value: Law Enforcement as Diplomacy

**Question: Is the USCG the U,S, lead agency for the Forum?**

Rear Admiral Day: *The Coast Guard has some unique capabilities there, because here's the bottom line is we're not viewed as a military organization. We're viewed as a law enforcement organization, which opens up a lot of doors, particularly for our access to China. We have very good access to China, and the reason being is, is because we're viewed as a law enforcement, search-and-rescue, fisheries protection, environmental type of organization. We're not viewed as a military organization. The Coast Guard probably has better access to China than most any other government agency.*

*This distinction between military and law enforcement is crucial to understanding the Forum's success. Coast guard cooperation focuses on practical problems, illegal fishing, maritime safety, environmental protection, that all nations share an interest in solving.*

**Question: Do other agencies sit in the meetings?**

*Rear Admiral Day: State has been interested and follows our activity but have not directly participated because they are spread very, very thin. So they generally don't attend any of the North Pacific Coast Guard forum meetings. They're aware of activities and we report our activities to them. But are they directly engaged? No. We certainly keep other U.S. agencies interested in the North Pacific advised of our activities and the opportunities generated. But the point is that it's a significant strategic asset for the United States as we try to deal with Pacific issues.*

*Again, let's go back to the Arctic. I see sometime in the next decade, because again, of the significant amount of traffic that is going to start going up there and there will need to be a cooperative vessel traffic agreement for the Bering Straits, all access to the Arctic from the West is via the Bering Strait, between the United States and Russia. There will be persistent surveillance and shaping designated traffic lanes. The Forum has provided the foundation from which to shape such solution sets to new problems in the Pacific.*

*The North Pacific Coast Guard Forum represents a different model of international cooperation—one built not on formal treaties and political declarations, but on practical problem-solving and operational collaboration. By bringing together the coast guards of six Pacific nations with very different political systems and strategic interests, the Forum has created a platform for addressing common maritime challenges. As new challenges emerge in the Pacific, the relationships and procedures developed through the Forum will become increasingly important to maritime stability in the region.*

By Robbin Laird and Ed Gilbert
April 2, 2011

# Note: The North Pacific Coast Guard Forum still operates today in spite of global tensions or perhaps because of them.

In an era marked by intensifying major power competition and strategic rivalry across the Indo-Pacific, one multilateral institution continues to operate with remarkable consistency: the North Pacific Coast Guard Forum. Now in its twenty-fifth year of operation, the Forum demonstrates how functional cooperation can persist even as broader geopolitical tensions threaten to fracture regional security architecture.

The Forum brings together the coast guard services of six Pacific rim nations, Canada, China, Japan, South Korea, Russia, and the United States, in what has become the most durable maritime law enforcement cooperation mechanism in the region. Its longevity is particularly striking given that its membership includes nations with competing strategic interests and, in some cases, active territorial disputes. Yet the Forum continues to meet regularly at both expert and senior-official levels, with all six original member services maintaining active participation.

The Forum's recent trajectory illustrates both its vulnerability to geopolitical headwinds and its capacity to recover operational momentum. After 2019, face-to-face gatherings became increasingly difficult as broader U.S.-China tensions, the COVID-19 pandemic, and Russia's international isolation following its invasion of Ukraine created significant obstacles to multilateral engagement.

For several years, the Forum operated at reduced capacity, with member services maintaining minimal contact and suspending joint activities.

The resumption of full in-person meetings in 2025 marks a significant revival. In April, the China Coast Guard hosted the 25th North Pacific Coast Guard Forum Experts Meeting in Nanjing. The first time all six member delegations had gathered together since

2019. This was followed in September by the 25th senior officials' meeting in Shanghai, which brought together more than 80 representatives from the participating coast guards plus Chinese maritime authorities. The Shanghai meeting produced both a Chair's Statement and a concrete cooperation plan for the coming year, signaling renewed commitment to sustained engagement.

What explains the Forum's durability?

The answer lies in its relentless focus on operational necessity rather than strategic alignment. The North Pacific presents shared challenges that transcend geopolitical competition: illegal fishing operations that deplete common resources, transnational criminal networks trafficking drugs and people across maritime borders, environmental disasters requiring coordinated response, and the fundamental requirement for maritime domain awareness in one of the world's busiest ocean regions.

The Forum's agenda has remained remarkably consistent since its establishment: fisheries enforcement, maritime security, counter-illegal trafficking, illegal migration prevention, information exchange, joint operations planning, and emergency response coordination.

These are not aspirational goals but daily operational requirements for coast guard services operating in overlapping jurisdictions. A fishing vessel engaged in illegal operations in international waters does not respect strategic rivalries; interdicting it requires information sharing and coordinated action regardless of broader diplomatic tensions.

Current discussions emphasize several priority areas that demonstrate this operational focus. Combating transnational maritime crimes, particularly illegal migration and drug trafficking, requires sustained intelligence cooperation and coordinated enforcement action. High-seas fisheries enforcement demands synchronized patrol schedules and shared vessel tracking to prevent illegal operators from exploiting gaps in coverage. Strengthening information-sharing mechanisms allows coast guards to build common operating pictures despite limited individual surveillance capabilities.

The Forum maintains a pragmatic institutional structure built around rotating hosts and multiple specialized working groups. These include dedicated groups for Combating Illegal Trafficking, Fisheries Enforcement, Maritime Security, Emergency Response, Information Exchange, and Combined Operations, along with a Secretariat function. China currently chairs the Secretariat and leads the trafficking working group, demonstrating that leadership roles rotate among members based on agreed schedules rather than geopolitical weight.

This working group structure allows the Forum to compartmentalize cooperation, enabling progress in less controversial areas even when broader tensions limit engagement elsewhere. Coast guard officers can collaborate on search and rescue protocols or illegal fishing enforcement patterns without requiring resolution of territorial disputes or strategic disagreements. The technical and operational nature of coast guard work creates natural boundaries around cooperation that military-to-military engagement often lacks.

The Forum continues to plan joint exercises and combined operations, though implementation depends on broader political conditions. Capacity-building activities allow members to share best practices and technical expertise in areas like maritime law enforcement, pollution response, and search and rescue coordination. Participants consistently describe the Forum as a key regional mechanism for maritime security and governance in the North Pacific, suggesting genuine value beyond diplomatic theater.

The Forum's persistence offers important insights into the architecture of regional security cooperation in an era of strategic competition.

- First, it demonstrates that functional cooperation can survive even severe geopolitical strain when it addresses concrete operational requirements. Coast guard services need to coordinate activities in shared maritime spaces regardless of broader national strategic positions, creating persistent demand for cooperation mechanisms.

- Second, the Forum illustrates the importance of institutional design for resilience. Its focus on technical cooperation rather than strategic coordination, its distributed leadership structure, and its compartmentalized working groups all help insulate cooperation from broader political disruption. When tensions rise, the Forum can contract its activities without collapsing entirely, then expand again when conditions improve.
- Third, the Forum highlights the limitations of purely operational cooperation. It facilitates coordination on law enforcement and safety issues but cannot resolve underlying territorial disputes or strategic competition. Coast guards can share information about illegal fishing vessels while their governments contest maritime boundaries. This creates value but should not be mistaken for strategic reconciliation.

The Forum also raises questions about the relationship between operational cooperation and strategic competition.

- Does sustained coast guard engagement help manage tensions by maintaining communication channels and building interpersonal relationships?
- Or does it risk legitimizing behavior in contested waters by normalizing presence and operations?

The answer likely varies by context and participant, but the question remains salient as regional competition intensifies.

The North Pacific Coast Guard Forum's twenty-five year track record demonstrates that multilateral cooperation mechanisms can persist through significant geopolitical turbulence when built on genuine operational necessity. As strategic competition in the Indo-Pacific continues to intensify, such mechanisms become simultaneously more valuable and more difficult to sustain.

The Forum's future will likely follow the pattern established over the past several years: periods of robust cooperation when broader political conditions allow, punctuated by intervals of reduced activity when geopolitical tensions constrain engagement. What matters is that the institutional infrastructure persists, allowing rapid resumption of cooperation when opportunities arise.

For policymakers and analysts, the Forum offers a case study in how to design cooperation mechanisms that can weather strategic competition. Its focus on technical operations, distributed leadership, and compartmentalized cooperation creates resilience.

Its practical value to participating agencies creates persistent demand for engagement even under political pressure. And its careful avoidance of broader strategic issues allows cooperation to continue in its limited but important domain.

In contested waters, operational pragmatism endures.

# PART 2

# THE PERSPECTIVES OF THE DISTRICT COMMANDERS

The District Commanders of the United States Coast Guard (USCG) play a pivotal role in overseeing maritime safety, security, and environmental protection across regions as varied as the Great Lakes, and the Gulf Coast.

Their strategic perspectives and operational challenges reflect a microcosm of broader Coast Guard realities: a vast and growing mission set, persistent resource shortfalls, and the need to adapt to rapidly evolving threats and technologies.

This part of the book introduces the issues discussed in our visits to the commands in 2010 and 2011.

# 1

# THE USCG IN THE
# GREAT LAKES REGION

WE SAT down in mid-March 2010 with Rear Admiral Peter Neffenger, District Commander for the Ninth Coast Guard District, which has oversight of the Great Lakes. The Admiral discussed the role of the USCG in the Great Lakes region and the challenges to securing maritime commerce in the region.

**Question: How would you describe the Great Lakes region and the USCG role in that region?**

Admiral Neffenger: *Our role here is much the same as for the Coast Guard elsewhere, with strong support for maritime safety, security, and pollution prevention and response programs and operations. I'll amplify these later and discuss some of our unique roles here, such as significant icebreaking to keep the marine highways open and programs to guard against invasive species to protect our waters in close coopera-tion with our partners in Canada.*

*What makes the Great Lakes interesting are a number of things. First, just to set the scene, you have internal domestic waters that are shared by two sovereign nations, that in and of itself is fascinating because international maritime law does not typically apply here.*

*Instead, there are a number of binational maritime agreements that we use to help govern maritime operations on the Great Lakes. These cover search and rescue, cross-border law enforcement, aids to navigation, icebreaking, pollution prevention and cleanup, and the like.*

*The question is, how do we jointly manage these shared domestic waterways? Many vessels transiting the Great Lakes and St. Lawrence Seaway cross the border many times in the course of moving through the system, so a joint approach with Canada is key.*

*By agreement, we incorporate international treaties for purposes of the Great Lakes, but they don't necessarily have to apply. So that makes it fascinating because it's a region where you don't have the weight of international agreement to drive you to action, but you have the demands of commerce on both sides, along with the day-to-day reality of operating on these waters that force the action.*

*The other thing that makes it interesting is that you have eight states and two Canadian provinces, each of which has jurisdiction out to their respective pieces of the international border.*

*So unlike a coastal state, which has jurisdiction that ends typically three miles out and then everything else from three to twelve miles is U.S .federal jurisdiction, and beyond that is the high seas and EEZ and it's not like that here.*

*You have a domain in which there are many governmental actors present with quite a bit of interaction amongst all those actors. And just as an example, one case where that is posing some significant concerns and problems for both sides is the management of ballast water. I don't know how familiar you are with the ballast water issue, but invasive species is a huge issue in the Great Lakes. In fact, it's probably the number one issue here.*

## Question: Ballast water is the water dumped from ships?

Admiral Neffenger: *Exactly. Foreign oceangoing vessels have been coming into the Great Lakes through the St. Lawrence Seaway since 1959. Many of them carry freshwater ballast from overseas, discharge their ballast in the lakes when they take on cargo, and leave. Some 70 or so invasive species out of 180 identified so far in the Great Lakes have been tied to past discharges of ballast water from foreign ships.*

*For some time now we've been concerned—actually, 1993 is when the Coast Guard first started looking at the management of ballast water. By management we mean looking to see what's being discharged into the lakes. Since 2008, we've done 100 percent examination, jointly with Canada up in Montreal, of all ballast tanks of every so-called saltie entering the Great Lakes system.* We check for evidence of open ocean exchange of that freshwater ballast with ocean salt water.*

*We look for salinity in the ballast tanks. The idea being that if you do an open ocean exchange of ballast water, it should kill any freshwater organisms that might be there. Science says it removes about 95 to 98 percent of freshwater organisms.*

*Long story short, since 2008 there has been no unmanaged ballast water entering the Great Lakes system or that is, all ballast water brought into the Great Lakes has either been exchanged with open ocean salt water or is retained on board*

---

* "Saltie" is what the Great Lakes maritime industry calls the vessels coming from overseas.

*the ship. As I noted, there's a great deal of concern about foreign species being put into the lakes.*

## Question: So it's from species from seagoing vessels?

Admiral Neffenger: *Exactly, coming from seagoing vessels. There's also some question about inter-lake movement, but the real issue is the potential for invasive species to come in from overseas. Some groups have even proposed extreme measures, including closing the seaway and the various connecting canals and locks to oceangoing vessels from overseas. I think that this is unrealistic and unnecessary, but it illustrates that this is a very big issue.*

*As an aside, it's important to note that there are two approaches to ballast water management. One is salt water exchange, as I just mentioned; the other is to treat the ballast water before or during discharge to remove any harmful organisms. Ballast water treatment is the approach taken recently by the International Maritime Organization (IMO). Treatment is also the approach the Coast Guard has taken in our proposed regulations for ballast water management, which we published in late August 2009.*

*This is a proposed rule, not yet in effect, that would require ships to treat ballast water under a two-phased approach. The first phase would require ships to meet the IMO concentration-based standard for treatment initially, then would require a much stricter concentration-based standard in phase two following a practicability review of the current state of technology.*

*Meanwhile, in the absence of new federal legislation and while awaiting implementation of the proposed Coast Guard regulations governing ballast water management, a number of states have taken their own legislative action.*

## Question: In place of federal legislation for the ballast water problem?

Admiral Neffenger: *Right, for the ballast water problem. The states have begun to take action. And so you have right now several states that have passed ballast water management legislation, and other states like New York have set their own standards using authority derived from the Clean Water Act.*

*If you think about the way in which the lakes are organized, New York is the first state you pass through in the course of coming into the Great Lakes system. When New York passed its own ballast water discharge standard, it effectively established a requirement that affects the entire system and notably Canada.*

*The New York standard references the International Maritime Organization standard for treatment of ballast water, a standard we use in our proposed rule, but one that the U.S .hasn't yet ratified. which requires treatment to ensure no more than a specified number of certain-sized organisms per a given volume of ballast water are discharged.*

*But the State of New York standard requires treatment to a level from either 100 or 1,000 times that of the IMO standard, depending upon the date of construction of the ship. These levels are not currently achievable by existing technology.*

*Their standard requires any ship transiting their waters to have equipment installed that can treat to 100 times the IMO standard for existing ships by 2012, and 1,000 times the IMO standard for new ships by 2013.*

*That standard will apply to just about everybody who moves through, because just in the course of coming down the US portion of the St. Lawrence Seaway, you cross the border some two dozen times into New York waters.*

## Question: So New York's approach affects all ships?

Admiral Neffenger: *That's exactly right. They're effectively regulating the entire system and everybody upstream of New York, including Canada. They're even potentially regulating intra-Canada commerce. Ships transiting to a Canadian port in the Great Lakes would have to meet New York standards if they transit New York waters.*

*As I noted, you're dealing with a very different kind of system here. If a coastal state were to pass a standard like that, ultimately a shipper could decide just not to go to that coastal state.*

*It's very different when you have one state that can affect an entire maritime system. So one of the things that's fascinating about this region, and an under appreciated piece of what the Coast Guard does, is how much people turn to the U.S. Coast Guard for adjudication, negotiation, and arbitration of those issues because we own a large piece of the ballast water management issue.*

*The Coast Guard has been responsible for regulating the operational discharges from vessels for quite some time. We were specifically authorized to regulate ballast water under the National Invasive Species Act of 1996, and our proposed ballast water discharge standard will be promulgated under that Act.*

*Previously, the EPA specifically excluded operational discharges from mobile sources, such as vessels, from the Clean Water Act. Several environmental groups sought to remove that exemption over the years. In 2004, a federal judge ruled in favor of a petition to manage vessel operational discharges— bilge water, gray water, ballast water, and so forth—under the Clean Water Act.*

*In response to the court's ruling, the EPA created the Vessel General Permit (VGP) in the fall of 2008. It establishes standards or best management practices for vessel operational discharges. Of note, the VGP also allows individual states to set their own standards for their respective waters.*

*Section 401 of the Clean Water Act allows this "state certification," and most of the Great Lakes states have used this certification process to set their own standards. Some have set theirs at the International Maritime Organization standard, and some, such as New York, at levels 100 or even 1,000 times higher than IMO.*

*In Washington, DC, the Coast Guard and the EPA are negotiating on how these Vessel General Permits and state standards will be enforced. It's possible an MOU will be developed between the USCG and the EPA that will detail what the USCG will enforce or report in support of EPA's VGP requirements. But in general, the Coast Guard does not enforce state standards.*

*This ballast water challenge is probably the number one issue up here. Until recently, it wasn't getting a lot of attention because it's largely considered not anybody else's problem, but we are slowly moving toward a national policy on invasive species and ballast water by virtue of what's happening on the Great Lakes.*

**Comment: If we would take this ballast discharge issue, we could say on the one hand, this is a case study of the absence of international law or regulation, or a kind of anarchy, so to speak. a cacophony amongst different perspectives. One state is determined that it has the right to execute a version of the law itself. So that's one driver for change.**

**The other driver of change is that the EPA now feels that this should be subsumed under the Clean Water Act, but there's absolutely no regulatory mechanism to actually execute this, and presumably, as usual, you're**

**supposed to come up with the ability to do this with no additional funding.**

Admiral Neffenger: *That's right. We fund out of our budgets here a near-fulltime presence up in Montreal. It's a rotating presence on temporary assignment, but it's a near-fulltime presence in Montreal during the shipping season to work jointly with the Canadian government to go on board every vessel coming into the system to examine their ballast tanks and to look for evidence of salinity in ballast tanks. It's a fairly involved process. It's an effective process.*

**Question: Every vessel?**

Admiral Neffenger: *Every single vessel, 100 percent. Every vessel that comes into the system.*

**Question: And where is it inspected?**

Admiral Neffenger: *In Montreal. That's the first port of call, if you will. It's the first call-in point for a vessel coming into the Great Lakes system.*

**Comment: But the Canadians must be concerned about the State of New York hijacking federal authority. And must be reassured with the system which you are doing with regard to inspections.**

Admiral Neffenger: *The Canadians work with us as well because they see us as the potential arbiter of this system on the US side. In fact, we just recently had a visit from Canadian officials to talk about their concerns about ballast water, and they've asked us for our continued assistance in pushing for federal-level, binational management. They know that we favor a federal solution to the ballast water issue. I've said this in my comments to the Great Lakes Commission.*

**Question: What kind of money are we talking about? How much money do you have to allocate annually to do this kind of Montreal inspection? And I would underscore, this is not in your list of mission sets.**

Admiral Neffenger: *That's a good question. I'll have to get you the exact numbers on that, but it covers the cost of travel and housing for the Coast Guard members I send there. With respect to our missions, our work in Montreal is an outgrowth of our marine environmental protection mission.*

**Question: And the disruption of commerce, which could occur if the New York law is enforced, would be significant, would it not?**

Admiral Neffenger: *If the New York issue or rather, if the question as to whether the federal government rule making will ultimately have primacy can't be resolved, it's likely to become one of the major factors determining whether ships will come into the Great Lakes system from overseas. It's also very likely that Canadian shippers will be faced with some tough decisions about moving cargo in and out of the central part of their country.*

*There are also concerns among some ship owners about the technological feasibility of meeting the New York standard by the deadline.*

*So it really is important, and there's a lot to say about the ability of the Great Lakes economy to survive if we don't collectively resolve this current issue with a single, understandable, and ultimately attainable federal standard for ballast water.*

**Comment: I'm old enough to remember the seaway opening, and so this is a huge infrastructure that we spent a lot of money building. Then you build communities around this, and then you're going to shut it down because of the desire of one state.**

Admiral Neffenger: *And of course, Canada is our largest trading partner by far. The two largest trading partners in the world are the U.S. and Canada, and a very large percentage of that trade goes via the Great Lakes system.*

**Question: And the cost of doing the ballast water treatment where are the funds for this going to come from? Certainly not from the State of New York?**

Admiral Neffenger: *Much of the cost of meeting the ballast water standard will be borne by ship owners and operators. As for the cost to the Coast Guard, it will all depend upon the amount of additional inspection, examination, or certification we would be required to perform to ensure compliance.*

*To summarize, we do what the Coast Guard does in most places, with some unique differences. Also, like most of the Coast Guard, we do the very best we can with what we have. We are a very lean organization here and elsewhere. We have little bench strength, and when we have to surge for major operations, we are vulnerable with the gaps we leave behind.*

By Robbin Laird and Ed Gilbert
May 3, 2010

2

---

# THE USCG 8TH DISTRICT
# FACES THE FUTURE

IN EARLY APRIL 2010, we discussed the challenges facing the USCG 8th District in managing safety and security in the Gulf region.

The scope and size of the territory facing the men and women of the USCG serving in the 8th District is immense. The district covers all or part of 26 states. Although a single district, it has three distinct regions, from the Gulf to the Midwest.

Of course, the offshore oil drilling efforts of the Gulf operate within this district as well. Clearly, the challenges associated with oil drilling are going up, not down. This is especially so because the new offshore efforts will be further and deeper than the current offshore activities. It will be more akin to the North Sea than to current Gulf offshore drilling.

And the USCG will need additional resources for inspections, helicopters, and ships to monitor and to ensure compliance with the national effort to shape a deep-sea offshore oil and gas enterprise.

**Question: Could you describe the size and nature of the 8th District?**

Captain Arenstam: *The 8th covers all or parts of 26 states from Mexico up to Canada and from the Rockies to the Appalachians. Roughly 35 percent of the U.S. population, or 100 million people, reside within the boundaries of the 8th District. Our workforce here includes 4,000 active duty and reserve military forces, 300 civilian employees, and 6,000 auxiliaries. Unlike any of the other Coast Guard districts, the 8th District has three distinct geographic regions. To the north is the part of the 8th District we refer to as the Western Rivers or the Old 2nd District.*

*This region covers all or part of 20 states and contains over 10,000 miles of waterways including the Mississippi, Missouri, Illinois, Ohio, Tennessee, and Arkansas rivers, and all their tributaries. Up in this region there are three sectors: Sector Ohio Valley, Sector Upper Mississippi River, and Sector Lower Mississippi River. There are 10 MSUs, 18 river tenders, and two aids-to-navigation teams.*

*So if you think about going all the way up to the North Dakota floods, going all the way over to Pittsburgh, we've got Pendleton, West Virginia or that whole river system there. So basically, if it flows to the Gulf of Mexico, it's ours.*

**Comment: Not only do you have a big swath of territory, there are also a number of game changers coming into your waterways, ranging from large liquid natural gas ships to larger container ships coming through the enlarged Panama Canal.**

Admiral Landry: *These are major challenges, and part of how this challenge is being viewed is through the eyes of the major ports in the region. All the major ports are trying to sort through how to meet the challenges and to remain competitive as shipping companies look at the best ports through which to operate.*

*The other major challenge is the deep offshore oil and gas enterprise. The companies are going to be drilling and working so much farther offshore and deeper; the offshore energy sector is a huge game changer. Field technology-wise, what they're doing now pales in comparison to two years ago. They're not only going 10,000 feet deep; they're also drilling deeper than before.*

*Captain Arenstam: The next zone I will talk about is the coastal zone, which I call the traditional zone here on the coast. It includes all or part of six states and contains more than 8,000 miles of coastline and nearly 1,300 miles of the Intracoastal Waterway.*

*The 8th District ports in this region receive approximately 26,000 deep-draft vessels annually, which equates roughly to 27 percent of U.S. deep-draft arrivals and nearly 90,000 towing vessel transits per year. In New Orleans alone in 2008, there were 46,000 towing vessel transits.*

*The 8th District coastal region is divided into four sectors: To the west there is Sector Corpus Christi and Sector Houston-Galveston. Together they have six small boat stations. Air Station Corpus Christi operates as an integral part of Sector Corpus Christi while Air Station Houston operates under the Operational Control (OPCON) of D8.*

*To the east, there is Sector New Orleans and Sector Mobile. Together they have nine small boat stations. Air Station New Orleans operates under the OPCON of D8. Aviation Training Center (ATC) Mobile, which is a Coast Guard Headquarters unit, provides direct aviation support to D8. Collectively, among the four coastal sectors, there are 17 patrol boats, 9 aids-to-navigation cutters, and 13 ANTs (aids to navigation teams).*

*Finally, the unique feature of the 8th District is the offshore sector. It captures the intercontinental shelf oil and natural gas industry. Within this region there are 6,500 oil and gas wells; 4,000 oil or gas production platforms, over 800 of which have full-time crew support.*

*There are 116 mobile offshore drilling units, 51 of which are stacked, some which are crude and some are not. There are 30,000 workers offshore on any given day. This infrastructure accounts for 30 percent of our domestically produced oil and 23 percent of our domestically produced natural gas.*

Admiral Landry: *This is the only district in the country where there is an offshore zone. The district commander serves as the Federal Maritime Security Coordinator. Post-9/11, the FMSC role that you see in sectors, Federal Maritime Security Coordinator under the MTSA 2002, this is the only one where the offshore sector is handled from the district. We are concerned with the security of the LOOP.*

## Question: What is the LOOP?

Admiral Landry: *The Louisiana Offshore Oil Platform, which is a mooring for imported crude that gets pipelined to the United States.*

## Question: How do you address security in the offshore zone? I frankly think that it is a challenge beyond the resources of the USCG and perhaps the US government?

Admiral Landry: *First, with regard to search and rescue, it is absolutely a partnership with the offshore oil and gas industry. The private sectors have to have a certain inherent capability themselves on the offshore platforms, but they also allow us to use their helo ports on some of these larger platforms.*

*If we're flying a search and rescue mission from the air stations offshore to rescue somebody, whether it's on the plat-*

*form itself or if it's a commercial fisherman in the area or a recreational boater, we can refuel on their fuel ports and we can continue on with the mission. So it's very much a cooperative effort in terms of the safety piece.*

*That's one aspect. The other aspect obviously is the safety after storms, hurricanes, things like that—the safety of the platform itself; and we work with them. We do so through a Mineral Management Service or MMS protocol for that effort.*

*If it's security, it's the same thing. It's integrated; it's absolutely the inherent self-security that they provide because they're required after 9/11 as a facility to have a security plan. Then the security we provide is for high-threat scenarios. And frankly, it would take the whole government to do the work if we had a threat against the offshore, including NORTHCOM and DOD assets.*

Captain Arenstam: *It would be close to impossible to have enough Coast Guard assets to police this entire region given the resources we currently have available. We do not have a current requirement for a 1.0 coverage factor for patrolling this vicinity.*

## Question: Could we discuss further the deep-sea offshore oil and gas enterprise concept?

Captain Arenstam: *Future oil exploration, as the Admiral has mentioned, is well offshore in very deep water. There is recent discovery activity in the lower tertiary geological formations that extends to 175 miles from the nearest land in water depths greater than 9,000 feet.*

*Technology is new to the Gulf, like the Floating Production Storage and Offloading facility or the FPSO are being employed for production. The first one in the United States is here off the coast already. It's anchored off of Morgan City.*

**Question: What is an FPSO?**

Captain Arenstam: *It stands for Floating Production, Storage, and Offloading. It's common in the North Sea and other parts of the world, but the U.S. hasn't allowed it.*

**Question: So this is our first one?**

Admiral Landry: *The LOOP is just a gas station for importing crude. This is actually going to be not only an import but also a production facility.*

**Question: And this is parked where?**

Captain Arenstam: *Well, currently it's anchored off the coast of Louisiana, but it will be out offshore.*

**Question: So this is a new technology that is expanding. What you can say about this region is they continue to bring on new technologies and new systems to support energy both production and importation. So this will be a dramatic increase in the challenge which you will be facing in the years ahead?**

Captain Arenstam: *Correct. There are currently 10 additional FPSOs, which are floating production facilities at various stages of construction, headed to the Gulf of Mexico. This will bring the total to 48 by the end of 2010. Discoveries in the Walker Ridge area are thought to be approximate to the crude of Bay Fields as well. This is a whole new game.*

**Question: This will be a producing and processing facility. In contrast, the LOOP is a floating gas station?**

Captain Arenstam: *The LOOP is not really a producer but is the nation's only deep-water oil port. It's located 18 miles offshore. It receives roughly 13 percent of the nation's imported oil, approximately 1.2 million barrels per day.*

*This oil is pumped inland approximately 50 miles and stored in caverns with a total capacity of 50 million barrels where it connects by pipeline to roughly 50 percent of the U.S .refining capability.*

*The Coast Guard provides several aircraft overflights of the LOOP each week. We do cop-on-the-beat air patrols of the loop in the offshore district.*

**Question: The Gulf is a crucial contributor to the national economy. Can you give a picture of the contribution?**

Captain Arenstam: *The Gulf of Mexico itself plays a crucial role regarding the production and import of oil for our nation's energy needs. Twenty-five percent of the domestic oil production occurs in the Gulf, with 18 major platforms accounting for half of that total. The destruction of a single one of these platforms in the Gulf would negatively impact our nation's economy.*

*Twenty-five percent of the nation's oil imports occur in the Gulf through lightering operations off the coast. The lightering zones are established to accommodate the Very Large Crude Carriers and Ultra Large Crude Carriers. There are four designated lightering zones used by single-hull tankers until 2015, and then there are six traditional lightering areas for double-hull tankers. The largest and busiest in the US receives 30 percent of the nation's crude.*

**Question: Can the Gulf ports take the new double-hull tankers?**

Captain Arenstam: *You have ultra large crude oil tankers which cannot get into U.S. ports. These are drawing 50 to 60 feet of draft, which are never going to get into Houston.*

*Instead of delivering at a refinery, they have to do it offshore to smaller tankers like a shuttle service of oil. Sometimes they could put in a pipeline loop that goes through a pipeline, but sometimes they lighter the ships and the ships bring them in.*

*Admiral Landry: It is a new approach to the oil distribution system, and if you look at Singapore or ports around the world, they work containers the same way. Huge container ships offload containers so the offshore becomes a port in itself, a geographic port even though it's water, and that's how they do it in Asia to deal with the lack of land*

Admiral Landry during the interview.

USCG Captain John Arenstam, Waterways
Management (8th District).

By Robbin Laird and Ed Gilbert
May 6, 2010

# 3

# THE USCG NEW ORLEANS
# SECTOR FACES THE FUTURE

WE VISITED U.S. Coast Guard facilities in the New Orleans area in 2010 and attended the keel laying for the new Sentinel Class Patrol Boats at the Bollinger Yards. Unfortunately, recent developments in the Gulf would demonstrate once again the importance of the USCG to the nation.

This sector, along with other USCG sectors, is facing significant challenges from resource shortfalls. We discussed all of the various mission areas with our USCG interlocutors, but we have provided only some excerpts from the wide-ranging and frank interview with Captain Ed Stanton and Commander Stocklin.

Captain Ed Stanton, Captain of the Port of New Orleans, is currently also commander of the unified command center, which has been set up on the outskirts of Houma, south of New Orleans, to manage the oil spill response in the area. BP, the Coast Guard, and state and federal agencies have been gathering resources to deal with the crisis.

**Question: Could you explain the focus of your activity in the New Orleans sector?**

Captain Stanton: *Sectors are alike organizationally, but because of their geography, the work focus is unique to each sector. They're all very different. For example, Miami is heavily counter-drug, counter-migrant, and cruise ship. New Orleans is heavily maritime safety, outer continental shelf, and river. It's a 300-mile-long port. New York is the UN, a lot of airport security, 30 congressional districts, and not as much shipping activity.*

## Question: What is the major focus of Search and Rescue (SAR) in your sector?

Commander Stocklin: *With regard to SAR, there's a lot on Lake Pontchartrain. We've got a good location with Station New Orleans to cover that. Also, we've got a good deal of offshore SAR where we've got stations in Venice and Grand Isle to respond, but their boats can only go 30 nautical miles offshore safely for a SAR response. It's a bit more of a challenge on this far western side of the AOR, because our small boats from the stations don't have very long legs, and we're constrained with our 87-foot patrol boats.*

*The 87-foot boats are assets that are so handy for us because they work so well in a variety of mission areas. We tend to overuse them, quite frankly, because we have to. For SAR where they're outside of the station coverage, they're the only thing we've got other than aircraft.*

*For security activities, any time it's rough going out the mouth of the river, we have to use the cutters because the station boats are too small to get there.*

*Search and rescue is an area where we work particularly closely with a lot of our state and local partners. State wildlife and fisheries, local sheriff's departments: they really collaborate a lot with us on the search and rescue mission.*

Captain Stanton: *Also, SAR for us really covers the wetlands. We have a lot of wetlands and estuary and a lot of shallow water. Forty-ones and twenty-fives (boats) can't go to most of the areas where we have SARs. So the air station becomes a critical piece; helicopters are especially important, as are the local agencies because they have airboats.*

*We rely heavily on the sheriff's departments, Fish and Wildlife Service, both from Louisiana Wildlife, Department of Wildlife Fisheries, and the national Fish and Wildlife folks for their airboats. The NOAA surveys on this area for coastal surveys are over 100 years old. So if there's not a depth marked on the chart, we're not by policy even supposed to go there. We will if it's critical and no other asset is available, but a lot of this is just too shallow for our boats.*

*Helicopters are really important for search and rescue in this area. Two weeks ago we had a tanker offshore that lost propulsion and was loaded with crude oil. They were drifting down within four miles of one of the largest oil platforms in the Gulf, which was an access hub for about 20 different pipelines. We had nothing out there. There was no way we could get there. Luckily, the Gulf is full of offshore supply vessels, and one put a line on it and was able to nudge it far enough, change its vector, drift just enough to get it past the platform.*

*So we rely on luck and other people's resources all the time.*

**Question: But luck may be useful in getting a result, but not a good planning guide. Clearly, the new cargo and tanker ships are getting bigger, and the challenges grow with them. I see little national focus on this challenge associated with global economic growth.**

Captain Stanton: *You're right, the vessels are getting bigger. The activity is getting bigger because they have to go farther into the Gulf. That's where most of the active drilling is now. What has not been tapped yet is the natural gas resource in the Gulf because of the price. The price is depressed, so they're not doing any exploration or production on natural gas. But when that happens, we will see the much larger Liquefied Natural Gas (LNG) vessels populating the Gulf.*

**Question: Maritime safety is obviously a big part of your activity here as well. Could you describe your efforts in this area?**

Commander Stocklin: *Just the sheer volume is something to note here. We have over 5,200 vessel arrivals, these are deep draft vessels, every year. Ninety-three percent of that is foreign flag. So that equates to a very heavy port state control workload.*

*Our focus is to go on board and do all the safety inspections on foreign flag vessels. In addition, we have the security workload of vetting them beforehand, and we have to decide selectively who needs a security boarding. So safety and security are kind of a one-two punch for us.*

*In addition to the foreign flag deep drafts, we have over 2,500 domestic vessel inspections per year. That's everything from the offshore supply boats to barges carrying oil and hazardous material. So that's huge.*

*And we also have to do facility inspections. Those are our refineries, chemical plants, anything that's required to have a security plan by MTSA. Not only is the number itself large, but we have some of the largest refineries and chemical plants in the country on this stretch of river. We approve their security plans, and we have to actually do on-site inspections. I think it's at least once a year, and additionally we're out there doing spot checks on a random basis.*

**Comment: Given your limited resources and the significant demand, I cannot imagine you can do anything other than a risk-based approach to security management.**

Captain Stanton: *We try to take a risk-based, strategic approach. And we work significantly with partners in the private sector as well as the local law enforcement sector. We look for things like cameras that can be shared on websites up and down the river in a regional security network. We look for the ports themselves to buy security boats that can be used up and down the river.*

**Comment: Your inland waterways challenge seems complicated by the age and nature of the boats you have available to do the missions.**

Captain Stanton: *From the mouth of the river to Baton Rouge, I don't believe there is a single boat ramp or a marina that dispenses gasoline. So our 25-foot boats are completely inappropriate for the tasks simply because they can't go very far up the river.*

*We use them right here in the city because they come from the lake [Lake Pontchartrain]. It takes them an hour and a half to get from my station on the lake to the river before the security escort on a cruise ship, for instance. And while they're doing that, it's a two-boat requirement, so there's no SAR coverage on Lake Pontchartrain, and the USCG Auxiliary must step up and take over the SAR.*

*We try to leverage everybody because of resource constraints. We have police boats and police authority, state police, the Louisiana Department of Wildlife and Fisheries, and the sheriff's department to help us out. We could not function effectively without them.*

USCG Captain Ed Stanton.

Commander Stocklin

By Robbin Laird and Ed Gilbert
May 7, 2010

# 4

# AIR STATION MIAMI: THE USCG IN THE FIRST LINE OF SAFETY AND SECURITY IN THE CARIBBEAN

WE SPOKE in 2010 with the USCG Miami Air Station Commander and staff about their area of responsibility, their operational concepts, and the upcoming addition of the Coast Guard's latest aircraft, the Ocean Sentry.

The Air Station operates in a demanding AOR characterized by significant maritime and air traffic that shapes commerce, law enforcement, and environmental challenges. Managing illegal immigration and drug trafficking remains a central concern, and because many migrants arrive from regions outside the immediate area—including zones with known terrorist activity—even routine immigration issues carry national security implications.

The interview addressed a broad range of topics, but the remarks selected here focus primarily on aircraft operations in the region and how the new aircraft will reshape capabilities in the AOR.

**Question: How many aircraft do you currently have at the air station?**

Captain Richard Kenin, Commanding Officer, USCG Air Station Miami:

*Right now, we have 10 aircraft. When I was stationed here in 1991, we had 21 aircraft at this unit. We're now down to 10.*

## Question: You had 21?

Captain Kenin: *We had 21 aircraft here in 1991 when I was previously stationed here, and we now have 10.*

**Question:** And presumably, the geographical area hasn't changed?

Captain Kenin: *No. What changed is that the interdiction mission shifted to Jacksonville, and our helicopters were sent away. But those helicopters did far more than just drug interdiction. They supported all the other missions for the air station. So the station was really downsized and re-designated in favor of standing up a single-mission unit.*

*We're phasing out the Falcons in favor of the Ocean Sentry. Our first Falcon has already departed, so we now have five of these aircraft, and as the HC-144s come online, the remaining Falcons will leave.*

*Notice up there we have 430 days away from home station (DAHS). We have an aircraft deployed 24/7 somewhere in the Caribbean. So we typically have four aircraft here, and on a good day, three of those are available for flying. We'll pick up one more.*

*The 144 you saw outside is coming online. We're predicting initial operational capability with that aircraft on October 1st, which means we'll be flying both the Falcon and the HC-144 together during our ready search and rescue missions as well as deployments. By October 1st, we'll have three aircraft. By next October, we'll have only 144s. The long-range plan, as you probably know, calls for seven of these aircraft here. Some are now saying five aircraft. Others suggest it might be as few as three.*

## Question: What are the major threats you face in your AOR?

Captain Kenin: *We have to cover a large and busy AOR with these aircraft and face a diversity of threats. Probably the biggest threat right now is migrants. And the threat isn't the Cuban migrant coming from Cuba, they come across for economic reasons, which isn't really a threat. The real threat from a Homeland Security perspective is people coming through the Bahamas. It's very easy to enter the Bahamas, and then it's only 40 miles from there to the US. We've picked up boatloads of Sri Lankans, Nigerians, Pakistanis, people who represent the national security threat we're truly concerned about. From Bimini, that 40 miles takes just an hour and a half. Obviously, there's also a serious threat from a search and rescue perspective.*

## Question: Obviously, a key element of addressing this challenge is providing ISR to locate threats and then having surface assets to prosecute them. How will you shape your ISR mission with the new aircraft?

Captain Kenin: *The HU-25, when it came online in 1980, was designed for a mission we no longer have. At that time, the thinking was that we would take the search out of search and rescue. We needed an aircraft that could quickly locate a vessel in distress, and then helicopters would extract people from the water. It was designed for that mission and performed it very well. It had excellent sprint capacity for distant search and rescue operations, great dash speed, designed for minimal loiter time before returning.*

**Question: And presumably, you're saving the cost of a surface ship by using this aircraft?**

Captain Kenin: *Exactly. You had a plane taking the search out of search and rescue. The Coast Guard air force does it all. That was the leadership's thinking in pushing us toward a jet, and it performed that mission very well. It also excelled at air interdiction when we began targeting drug smugglers, particularly with a different radar configuration.*

*But now the Coast Guard's mission for fixed-wing aircraft has evolved. We're now focused on maritime patrol, and that aircraft doesn't have the endurance. It has the legs but not the staying power for the new Coast Guard mission. We need aircraft with sophisticated sensor packages. Over the years, we've added various capabilities to that aircraft, making it increasingly complex. The avionics systems became more complicated, requiring more maintenance and care, with numerous add-on capabilities.*

**Question: But these capabilities aren't integrated?**

Captain Kenin: *They're not integrated. We've hung more things on them. Individually, these packages are good, it's a good radar, it's a good FLIR, but they don't integrate well, and the aircraft simply can't stay out there long enough to accomplish the mission.*

*When we locate a go-fast moving across from somewhere, we can't stay on scene long enough for the surface fleet to react and make the interdiction. We can find it, and we do that well. But if we've already been flying for two and a half hours, all we can do is report a position and return for fuel.*

*The aircraft was optimized for a particular mission set, and now you're talking about a multi-mission requirement. The aircraft might launch with a specific responsibility, search and rescue or ISR in support of drug interdiction. but when it's out there, it shouldn't have to return to transition to the next mission. It's there, and you can shift it seamlessly. If there's an emergency search and rescue elsewhere, it should be able to support that. It should be passed functionally from one mission to another.*

**Question: Presumably multi-mission in terms of not just ISR but also lift and carry?**

Captain Kenin: *Yes. That provides tremendous flexibility given that we're a smaller air force. Multi-mission is huge, especially in this AOR, because there are so many different places you go where you need to shift tasks in flight.*

**Question: We assume that as you stand up the new aircraft, it will take time to evolve the capabilities and your operational concepts for employing them. It will be important to build in tolerance for evolution or tolerance for change in shaping capabilities over time. Is that something you're concerned about, having the tolerance for capability evolution rather than expecting perfection from the start?**

Captain Kenin: *You're right. We need to build space for evolution. We have a pretty good product, but not where we need to be from the get-go. The sensors themselves are solid. The aircraft has good radar that's done well in testing. But the complaint is that it's difficult for operators to use. You want something that works well, but as you point out, can the operators use it effectively?*

*The most expensive sensor in that airplane is the pilot. We've invested enormous resources training these people, the sensor operators. In any organization, any airline, it's the pilots we train, the sensor operators. Give them tools that actually work.*

*That's what we're dealing with regarding the pilot interface. The concept is excellent: it's multi-mission. We paid a premium for it to be multi-mission. If you look at aircraft like the Dash-8s or P-3s, there's no ramp, no way to use them for logistics. It's easier for them because they load sensor equipment in the back and focus entirely on sensor missions. This airplane has a ramp. I can throw a rotor blade in the back. I can do many different things. I think it's an excellent fit for what we do here because when you look at the AOR, there are numerous short runways we couldn't access in the Falcon. Now we can. They're all available to us.*

**Question: The CASA aircraft was designed partly to operate in the Mediterranean, and the Caribbean has considerable similarity. I would assume the fit is quite good?**

Captain Kenin: *It is, and we need the loiter time this aircraft provides. We really need an airplane that can stay out seven or eight hours, and that's important for several reasons. First, you can actually cover the ground you need to. Second, it takes considerable time for a ship to arrive. We've had problems in the past where I found the target but had to leave, and we lost it. By the time the ship or another asset arrives, whether CBP or another Coast Guard asset. It's difficult to relocate. They've disappeared. We play considerable cat and mouse with smugglers in the Caribbean, so being able to maintain contact is crucial because they'll try to dodge and hide until we're forced to depart.*

**Question: So the multi-mission capabilities of the aircraft coupled with its loiter time fit your AOR and multi-mission operational concepts quite well?**

Captain Kenin: *We believe so. Our resources are multi-mission, our people are multi-mission. The Navy, DOD, and other services have the luxury of specialized personnel. But our people are doing ISR and then running up front to drop a pump, going back to fill the aircraft, doing all these different things.*

**Question: How would you describe the challenge moving forward?**

Captain Kenin: *We've been given a new aircraft, and everybody's excited. We all agree this will provide significantly more capability than the Falcon. But how do we mold and change our operations to continue meeting operational commander expectations with this new aircraft?*

*That's something we're developing as we train on it. We'll learn how to use it, discover what the capabilities are, and leverage the ability to operate at much smaller runways, and that's enormous for us. We can go places we've never been able to reach before. So the question becomes: Is our mission always focused south to Cuba now? That's the question we're working through.*

The first Ocean Sentry MPA for the Miami Air
Station IOC October 2010

The Cockpit of the Ocean Sentry.

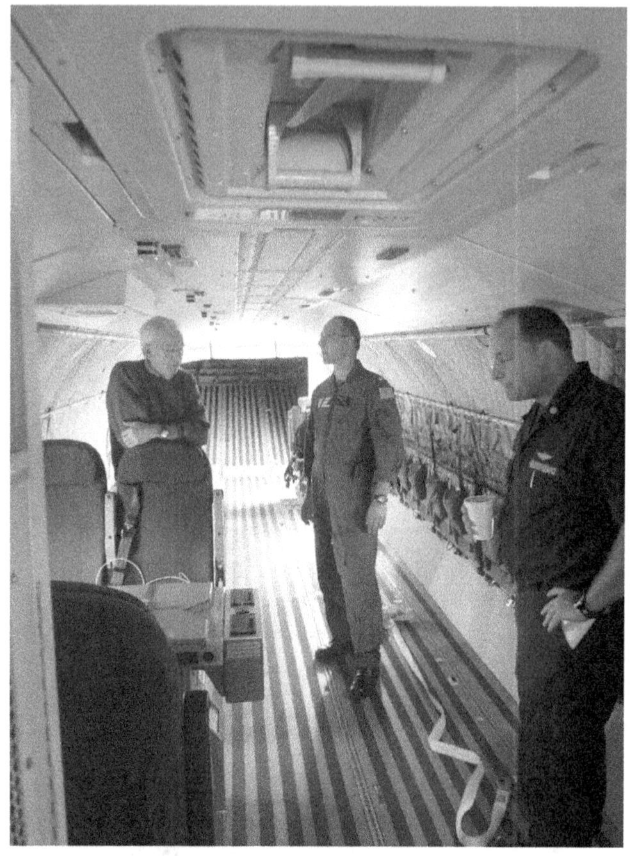

Inside the Ocean Sentry: "The Ocean Sentry
unlike the Falcon has significant cargo space."
Rear Admiral Gilbert is seen to the left in the
photo.

By Robbin Laird and Ed Gilbert
June 8, 2010

# 5

# THE VIEW FROM ELIZABETH CITY

WE WROTE three articles on the operations and challenges at Elizabeth City, North Carolina in 2010. These articles are all included in this chapter.

The primary U.S. Coast Guard facility at Elizabeth City, North Carolina is Coast Guard Base Elizabeth City, which encompasses several major commands and support units. The site includes Coast Guard Air Station Elizabeth City, one of the service's busiest aviation bases, alongside other key tenants such as the Aviation Technical Training Center (ATTC), the Aviation Logistics Center (ALC), and Station Elizabeth City, a small boat rescue station.

Major USCG Units in Elizabeth City

- Coast Guard Air Station Elizabeth City: Executes search and rescue, law enforcement, environmental protection, and logistical roles, maintaining a fleet of HC-130J aircraft and MH-60T helicopters.
- Aviation Technical Training Center (ATTC): Headquarters-level command that trains Coast Guard enlisted aviation personnel at multiple levels.

- Aviation Logistics Center (ALC): Responsible for logistics and depot-level maintenance support of all Coast Guard aviation units.
- Base Elizabeth City: Provides administration, logistics, engineering, medical/dental, and morale services for Coast Guard units throughout North Carolina and manages much of the installation's real estate and infrastructure.
- Station Elizabeth City: Operates small boat search and rescue missions as part of the broader mission set of Coast Guard Base Elizabeth City.

The base complex covers over 800 acres and includes runways, training facilities, logistics and maintenance hangars, medical clinics, and support buildings. It serves as a regional hub for Coast Guard aviation, maintenance, and technical training, as well as administrative and health care services for Coast Guard personnel and their families.

Elizabeth City's Coast Guard presence has strategic significance for USCG operations along the East Coast and supports both operational and technical missions vital to the service.

## The Challenge of Modernizing USCG Infrastructure: The Case of the Elizabeth City USCG Base

August 24, 2010

Earlier this summer, I interviewed Captain Bennett, base commander at the Elizabeth City Coast Guard Station. Captain Bennett discussed the challenges facing the facility, with aging infrastructure impacting operations and capabilities. Our website examines concepts of operations and capabilities. Clearly, one key aspect affecting both is the nature of the bases from which U.S. and allied forces operate. Con-ops and capabilities are clearly intertwined with the physical assets available to the forces.

But while there has been much debate about the platforms the USCG is acquiring to replace its aging fleets of aircraft and ships, little attention has been focused in the public debate on the base of the pyramid: the basing infrastructure.

**Comment: One of the things that people generally don't understand is the role of infrastructure and how its condition affects the operational capability of the Coast Guard. Give me a sense of how the condition of the infrastructure here on the base which is so central to the entire air capability of the USCG affects operations.**

Captain Bennett: *Essentially, this is a World War II vintage base. It has a deep legacy here within the community in Elizabeth City and has played a key role over the years since World War II. Really, any rescue that you see at sea or anything the Coast Guard is doing out there on the high seas or picking up people locally in lakes and inland waterways all that portion of operations can be reverse-engineered back to a base like Elizabeth City.*

*It is here where the folks go to do the mission, where they're trained, where they eat, where they sleep, and where the equipment is maintained and the training facilities are located.*

*So you can connect all the dots and go back to where our personnel are trained and where the equipment is maintained, and there's a key linkage there. If that foundation of support is not there, it's hard to get those Coast Guard men and women out there doing the mission with the right tools, the right training, at the right time.*

*It's something that the public at large doesn't see, but it's very important to our Coast Guard men and women.*

**Question: One of the things that's striking to me is that we rely on the Coast Guard to surge to deal with a national crisis like the Gulf oil spill. And we now have on this base the complete support element for the entire air capability of the Coast Guard.**

**Yet the warehouse that functions as the FedEx Memphis facility equivalent is in very old facilities that one certainly cannot call state-of-the-art. How important would it be to get more modern infrastructure to support the Coast Guard air fleet?**

Captain Bennett: *Absolutely essential. As you stated, we have a FedEx-style approach to aircraft maintenance. Elizabeth City maintains all the aircraft in the entire United States Coast Guard. This is the hub where it all happens, all the spare parts, all major maintenance comes through here, Elizabeth City, North Carolina.*

*Currently, our warehouse, which houses many of these spare parts, is in great need of repair. We have a crumbling floor right now that we're shoring up. This part of the country is close to the Dismal Swamp, and we have a lot of underground water. Actually, the warehouse sits on top of what amounts to an underground river.*

*So the floor is sagging, and we're looking forward to getting a new warehouse so we can adequately house all the spare parts with state-of-the-art warehousing.*

*Because if we have a failure here within Elizabeth City for aircraft maintenance, it will affect the whole fleet throughout the Coast Guard. We are a single point of failure. So that's a huge infrastructure issue that we're looking forward to working through.*

*But right now, we have a bridging strategy with a temporary fix on the floor, if you will.*

**Comment: Well, one way to look at the situation is that we've had a public debate for a long time about the aging fleet, the aging aircraft, the various aging assets, and the challenge of retaining Coast Guard personnel in difficult financial times.**

**But there's absolutely no public visibility regarding the state of the infrastructure, and I would look at the infrastructure as the base of the pyramid for modernization and recapitalization of the platforms for the USCG. But there seems to be little traction for that view. Give me a sense of the nature of this pyramid that's crucial to the actual capability of the Coast Guard.**

Captain Bennett: *Currently, this base, as I mentioned, is the hub of aircraft maintenance for the entire Coast Guard. One of the issues that the base commanding officer deals with and that I deal with is just the actual roadways here on the base. Basically, again, we're dealing with World War II-era roads.*

**Question: Quite literally?**

Captain Bennett: *Yes, literally. We have a lot of traffic, almost 3,000 people coming in and out of the base every day, between the folks who work here and those who come in for their ID cards, medical appointments, pharmacy visits, or to use our small exchange. So we have tremendous use of our roadways.*

*As our trucks and deliveries come in, there's a lot of wear and tear on the roadways. In our construction projects, we typically focus on buildings or air station hangars on the shore side. We never get down to the level where literally you're dealing with the ground. But it's a key issue. We have one main road here that runs along the Pasquotank River, and it's critical that we maintain it. Currently, my folks in facilities patch up the potholes as they can and keep things going.*

*I tell my staff we're like electricity: until you don't have it, you don't care about it. The roadways just don't get a lot of visibility. Similarly, starting from the ground and working up, we have many other buildings, but we can't house all of our outdoor equipment, our trucks and some of our yellow equipment.*

*It's all out in the weather right now. We'd like more garages for those. So all those things that are behind the scenes are very important to any municipality running a town or a base well. This base is really a small city, and I call myself the mayor. We focus on keeping the utilities operating, the electricity, the water, the roadways, all those things that are behind the scenes. They're behind all the rescues that you see at sea and all the Coast Guard men and women doing the operational mission.*

*One of the things in our tour we just looked at was our old gym. The gym used to be a chapel, and then it was a movie theater in World War II. Now we have aerobic equipment in there and free weights and whatnot. It's adequate and it works, but it's not the optimal solution.*

*We'll get a state-of-the-art gym and rescue swimmer training facility, but still, we have needs across the board for this base, state-of-the-art physical fitness equipment for all the folks who come through the base.*

**Question: And the point is you have 2,100 personnel, and they're not all going to be rescue swimmers?**

Captain Bennett: *Correct.*

**Comment: When you showed me all the outdoor storage, things that are exposed outdoors, and the World War II structures, one of the things that folks should realize is this is hurricane country. Given that**

**we've decided to consolidate all our aircraft support structure here, one thing that would concern me is the need to improve protection from the high winds and hurricanes in the area.**

**And so, it would probably be optimal if we had some new capacity here that was more state-of-the-art and also more hurricane-protective.**

Captain Bennett: *Yes, that's spot-on. We do have a lot of sheds, as you mentioned, and equipment outside that should be properly secured and housed in substantial buildings. One of my initiatives when I got here was to try to get rid of some of our sheds in case we have a big blow here.*

*We're working toward that, but it's been slow going because we've grown a lot faster than we've been able to build the infrastructure. So what we do when we have a threat of a hurricane is take some items inside and secure things as best we can. But it is a concern, because this is North Carolina, and we've been known to get some pretty big hurricanes through here.*

*Ironically, as we discussed, sometimes the aftermath of a hurricane will be so bad that we lose equipment and may have building damage. We do typically get some supplemental funding, which helps in the rebuild.*

*We were able to reconstitute our waterfront on the Pasquotank River after the last major hurricane we had. We purchased new riprap and reroofed some of our buildings.*

*So that's the plus and minus of hurricane season, but it's certainly not the way to hope for hurricanes or whatever to get money to help recapitalize your base.*

# Elizabeth City USCG Base Infrastructure

August 25, 2010

On my tour of the Elizabeth City USCG station there were several examples of the state of the infrastructure and of needs for modernization.

A first example is how much stuff is kept outside because of the absence of permanent storage facilities. This is obviously a challenge in a hurricane sensitive area.

A second example is simply the condition of the warehouse. If one remembers that this warehouse is the "Fed Ex" support center for the entire USCG fleet one gets a since of perhaps a weak link in the system.

A third example combines concerns about the condition of the warehouse with the lack of permanent storage space necessitating placing significant material outside of the warehouse.

A fourth example is the general condition of the roads and parking lots on the facility. Given that the roads are essential to warehouse functioning, the condition of these roads is an important factor as well.

The fifth, sixth and seventh examples underscore the World War II character of the base. The fifth example shows the parking lot, which relies on continued use of USN sea ramps from World War II.

The sixth example shows the recreation area, which operates on the ruins of the World War II USN base, which is now the USCG station.

The seventh example is the gym for the base. We expect our USCG staffs to be fit; they exercise in a World War II base movie theater. The theater is seen from the front and then the director for the exercise facility shows the projection windows, which still exist in the building from World War II.

The good news is that some modernization is afoot. Thanks to Kevin Costner and his USCG movie, the rescue swimmers of the USCG received notoriety. A new training facility is being built to support their training.

# ATTC : The First "One-Stop Shop" For USCG Aviation Assets Maintenance Training

May 7, 2010

In mid-March, I discussed the U.S. Coast Guard's evolving approach to training technicians and shaping an effective maintenance enterprise with the retiring (but only in the separation-from-service meaning of the word) Captain Joe Mihelic, Commanding Officer of the Aviation Technical Training Center (ATTC). Also contributing significantly to our understanding of the ATTC's role was the executive officer of the unit, LCDR Daniel Leary.

For a non-coastie, what is immediately noticeable is that the ATTC is co-located with the Aviation Logistics Center for the entire USCG. In the course of visiting both facilities, it was clear that the USCG benefits from the synergy between the two.

According to the ATTC website:

*The primary mission of ATTC is the safe and effective maintenance of the Coast Guard's aviation fleet. The quality of our graduates directly impacts the performance of our aviation resources today, tomorrow, and years into the future. This is a tremendous challenge to which we remain rigidly committed. To meet that challenge, we strive to ensure that each graduate meets the extensive requirements expected by the twenty-five U.S. Coast Guard aviation units across the country.*

**Question: Could you explain the role of ATTC?**

Captain Mihelic: *The purpose of the ATTC is to provide journeyman-level and advanced maintenance aviation enlisted maintenance technicians to the field units. Obviously, we have a very short period of time to take people who are green and get them out to the field, and give them the basic traits they need to perform in the operational Coast Guard.*

## Question: How long is the training?

Captain Mihelic: *For basic A-school, where we teach to the journeyman level, it takes:*

- *18 weeks for ASTs or Aviation Survival Technicians,*
- *20 weeks for our AMTs or Aviation Maintenance Technicians,*
- *and 20 weeks for our AETs, which are our Avionics Electrical Technicians.*

*So those are the three aviation rates that we have in the Coast Guard.*

*Right now, everybody who goes into Coast Guard aviation from a maintenance perspective on the enlisted side comes through ATTC. It's one-stop shopping. We are a small service: if you think about the 45,000 people in the Coast Guard, about 2,800 of those billets are in aviation, with 800 pilots and about 2,000 enlisted.*

## Question: So this facility trains for the whole Coast Guard?

Captain Mihelic: *This facility teaches all the aviation enlisted maintenance A-schools for the entire Coast Guard. Historically, at the advanced level or C-school level, the Coast Guard has focused primarily on rotary wing training and has done most of our investment in those training aids.*

*Over the last 10 to 15 years, as the model for ALC [Aviation Logistics Center] has changed to take on the depot-level work for all aviation assets, it's been a challenge to catch up and try to develop a fixed wing training capability, a proper fixed wing training capability with the same type of training aids that reflect the current fixed wing assets we have in the field.*

**Question: So that's the C-130?**

Captain Mihelic: *That's the C-130, the Casa, and the Falcon jet. Obviously, with acquisition reform, there's been a renewed emphasis on writing the requirements and getting the acquisition plans updated to get those types of training aids in the budget. Specifically on the Casa side of the house, we've done a lot of work in that regard to document those requirements and to get that acquisition plan updated.*

*These are long-term initiatives, nothing short-term, but in the long run, it will prevent us from having to do death by PowerPoint and then taking C-schools down to ALC to physically look at an aircraft that's going through the depot facility. That's what we're trying to avoid.*

**Question: How is training changing with the Casa aircraft coming in or the HC-130J coming in? Are there changes to how you're doing business?**

Captain Mihelic: *There are really two pieces to classroom training. You have initial theory and procedural training. Historically, here at ATTC we've presented a lot of material using PowerPoint. So we're trying to look at programs like Engrain and computer-based training that allows a 3D modeling perspective that is more interactive and much more realistic.*

*For the courses we currently teach, training is presented in the classroom; everybody comes to ATTC and the entire course is taught here. But the wave of the future may be to take a one-week course prior to getting here with a good computer-based e-learning environment to get prepped to come to school and then get into the meat of the course subjects.*

**Question: One aspect of the MPAs and the new Js is that you have more sophisticated, more complicated mission capabilities in both aircraft. How does that affect training?**

Captain Mihelic: *If you look at CG Modernization, it links all the training centers. There's a fine line between maintenance training and operator training. You see that more in the mission system pallet and the avionics-type gear. Our maintainers are our operators, and they need an understanding of both troubleshooting/repair and operation.*

*We really want ATC Mobile to be the lead and train on how to use the equipment from an operational standpoint and be very proficient in that. When the operators have problems, though, it's those same people that will troubleshoot and try to repair the equipment or work with contractors to swap boxes and resolve the software integration problems.*

*So collectively, we need to be linked closely with ATC Mobile. If they're giving operator training and they're seeing a specific type of problem, we have to incorporate that type of problem into our maintenance troubleshooting so that they're prepped to handle that when they see it out on an operational mission.*

**Question: What is your core focus within the training effort?**

Captain Mihelic: *We really focus on the maintenance aspect here at ATTC. The operational training would then be ATC Mobile. The one rate where we really cross boundaries is the*

*AST rate, providing initial Rescue Swimmer training. We teach the basic technical skills necessary to perform a water rescue, which is really operational-type training that we're doing here for the rate.*

*We instruct through scenario-based training. We give them the basic skills, both physical and mental, and provide a lot of training in the pool. The real test is a three-man scenario of a life-saving predicament where we have a member in a parachute, another person in a raft, and another person floating in the water.*

- *Can that rescue swimmer properly think and triage the situation and go for the person first who's floating facedown in the water?*
- *Then can he get the guy out of the parachute who's potentially entangled?*
- *And then can he handle the raft victim?*

*So we give them a very challenging situation with three different people in different situations, and the goal is to save them all. This really is an operational scenario, but we limit our operational training.*

**Question: So the technical training is embedded in the real operational scenarios?**

*Captain Mihelic: Correct. Now on the flipside, ATC Mobile owns the Standardization Team for the Rescue Swimmer Program. They're the ones that would go out and perform annual Stan Checks at our 25 operational air stations.*

*So if there's a Rescue Swimmer assigned, they'll be evaluated and given pointers on how to refine their skills and to make sure they're executing the mission and integrating with the rest of the crew and the pilots. ATC Mobile gets direct operational feedback to enhance their training. Any technical issues identified get referred back to us for review and potential response.*

**Question: You're close to retiring, and one question that comes naturally to mind is: what's the difference between the training that you got as a young aviator and a new aviator today?**

Captain Mihelic: *I started learning to fly in 1984. We did a lot of studying from books and then we jumped in an airplane. We did get some crude simulator time, but the advances in computer technology and computer-based training are light-years ahead of what we had to learn systems and practice flight scenarios. Today, we have tools to make flight handbooks come alive. In the old days, you just read and studied a lot.*

*At ATC Mobile, there have been tremendous advances in their simulator capabilities, especially visual simulation; it's much more realistic. We may not yet match some of the advanced simulation capabilities that some of the airlines have, where they can get people certified to land jets without actually flying in an aircraft, but we have made tremendous advances in that regard.*

**Comment: You mentioned earlier that in the acquisition reform, there's been an effort to highlight actual expenditures for training.**

Captain Mihelic: *Right. I'm more of a logistician. If you look at the 10 elements of logistics, training is an essential piece. I think we're doing a better job of documenting the actual requirements and the benefits of training and incorporating that into the budget cycle. It's easy to buy an aircraft and put it out in the fleet. It's another thing to make sure all the logistical support is there.*

**Comment: You have better recognition in logistics and maintenance. Both are an essential part of readiness, which enables operational capability.**

Captain Mihelic: *Definitely.*

**Question: And so you need training to have readiness. It's part of the cycle?**

Captain Mihelic: *You have to have a full logistics complement to make the asset work and sustain it so that we can effectively put steel on target.*

**Question: Final question: what investments would make important improvements on how the Coast Guard could operate more effectively?**

Captain Mihelic: *If we want to properly invest in training, we have to look at the current list of assets that we have out there operationally. We have to either invest in the infrastructure and the training aids to support those assets, or we have to potentially partner with either DOD or civilians to provide that training. The C-130J may be the best example.*

*We may never take J training in-house as long as there's a pipeline available to provide maintenance training. There is a multinational type of training arrangement that's contractor-based. In this case, it may not be cost-effective to pull that training in-house.*

*But if we're the sole user and there isn't any cost-effective commercial provider, we need to develop that organic capability.*

*The Casa may be a good example. We're implementing*

*new Casa courses in-house, as we operate a unique configuration of the airplane with the glass cockpit that we have.*

*We need to invest in those types of training aids and infrastructure so that we don't impact operational missions by trying to teach maintenance procedures on an operational asset.* *

---

* Casa refers to the CN-235 maritime patrol aircraft.

6

# THE STRATEGIC PERSPECTIVE FROM THE 11TH USCG DISTRICT

WE VISITED the 11th USCG District in July 2011. Rear Admiral Ed Gilbert was a former 11th USCG District commander. The current commander, Rear Admiral Joseph Castillo, provided a perspective on some of the key challenges posed by the size of the District as compared to his current capabilities.

The 11th USCG District is a very large area: The Eleventh Coast Guard District encompasses the states of California, Arizona, Nevada, and Utah; the coastal and offshore waters stretching more than a thousand miles from the California-Oregon border to the California-Mexico border; and the offshore waters of Mexico and Central America down to Ecuador. Coast Guard operational units are located throughout the state of California, with the Eleventh District and Pacific Area headquarters located on Coast Guard Island in Alameda, California, along the east side of San Francisco Bay.

Rear Admiral Castillo highlighted a key reality for the USCG that is oftentimes overlooked: the USCG has safety and security responsibilities for the inland waterways, a system which reaches significantly back from deepwater and port responsibilities, in addition to their more well-known coastal and open ocean responsibilities.

Indeed, one can conceive of ports as the hub of a wheel in which the spokes reach out into the ocean and back deep into the interior in certain areas.

Rear Admiral Castillo emphasized that with the very large area of offshore operations, it is critical to have enough assets to be physically present in the various diverse sub-regions throughout the AOR.

Rear Admiral Castillo: *You need to have enough assets so that you are actually near or physically present in an area of interest. With the vast area to be covered, and the fact that ships move relatively slowly, a significant number of assets are needed to maintain this presence due to the simple reality of time-speed-distance.*

*And modern and capable ships, to me, are at the heart of persistent presence in terms of ability to execute the mission in the offshore area. For example, if you are pursuing a go-fast vessel, you can have an MPA (Maritime Patrol Aircraft) tracking them. That is helpful, but I will need a ship partner with the MPA that is able to intercept, stop, and board that ship to fully execute the law enforcement mission.*

*What I need is a persistent presence that can not only survey the area and identify the traffic in that area but also be able to sort and target from among that traffic to identify those threats or challenges that are of interest to us.*

*Given the reality of the time/speed/distance problem, I'm just not convinced that you can achieve mission effectiveness without having a large number of assets, or a smaller number of more capable assets, that you can put against the threat or mission. And until we figure out how we can Star Trek transport things around, I'm still stuck with the physics of being able to get boots on target right now.*

*Will there be a way of delivering by aircraft, long distances at sea, to get somebody on the target of interest? That may be where we can go in the future as opposed to only vessels. Who knows?*

*What I know now is that I need to be able to operate in all weather conditions with significant sea states and for extended periods of time and over long distances, and be able to deliver helicopters and small boats to effect the end game, be it in law enforcement, search and rescue, migrant interdiction, or other missions. I need to be able to sort, and figure out whom it is that I need to target, and be able to get somebody on that vessel who can take law enforcement interdiction action and bring it to a successful conclusion.*

*When it comes to offshore search and rescue, when there's no Coast Guard vessel anywhere nearby, AMVER has been very successful and saved a number of lives. This is a voluntary merchant vessel tracking system operated by the Coast Guard where merchant ships share their transit plan and make themselves available to assist fellow mariners in distress in their vicinity. This partnership has been very helpful in identifying merchant vessels that can help out when we can't get one of our own platforms there.*

*We've also developed many great partnerships with other countries in the search and rescue realm as well. But we don't have the same number of search and rescue cases well off our shores and deep down in the Eastern Pacific as we do in the counter-drug arena.*

*In the Eleventh District, the counter-drug mission is the primary driver of the need for more ships to be available to prosecute, to evaluate, and engage whatever targets of interest we find.*

Another key challenge, which the Rear Admiral discussed, was managing the conveyor belt of goods coming into the United States.

Rear Admiral Castillo: *Almost half of all U.S. containerized cargo comes through the Port of Los Angeles/Long Beach. And if something were to shut that port down, the remaining ports on the west coast would not be able to handle the volume that one port complex handles. We work very closely with all of our partners in LA/LB, and in all our ports, to ensure the safety, security, and efficiency of the port complex.*

*And the process currently in place throughout the country requiring an Advance Notice of Arrival 96 hours prior to entering a U.S. port is critical to our efficient management of this "conveyor belt."*

*You never know for certain what's in a given container, and you've got to be able to reach back overseas and make sure that they've got the security in place where that container's first getting loaded on a vessel. To enhance the probability that a container will not be a Trojan Horse, you need to work with foreign authorities to reduce the risk inherent in the conveyor belt of global trade.*

*With the establishment of international standards, being able to confidently determine what's getting loaded onto vessels in foreign ports, and securing the cargo that's being brought into our port long before it gets here, that's a big part of risk management and defense in depth.*

**Question: What is the impact of the shortfalls in numbers of platforms on your thinking about operations?**

Rear Admiral Castillo: *We can't be everywhere for everybody all the time. We have some very capable assets, and some very capable people.*

*For a short period, we can apply a good number of resources to a specific problem when it happens. But the consequence is that we are not able to apply the resources we would want against our other mission requirements. The vast majority of our assets are multi-mission, but you can't necessarily use those multi-mission capabilities if you are overwhelmed by resource needs for a certain situation or location.*

*We're always ready to tackle something. We're not always ready to do every single thing at the same time.*

## Question: If USCG resources were cut an additional 30%, what would be the impact on operations?

Rear Admiral Castillo: *I think that from a policy standpoint, we would become very "near border" centric. And because that would affect the capacity to handle, even for a short period of time, that significant event that happens close to home, you would not see a lot of Coast Guard units forward deployed, in my opinion. If you're talking about the type of resourcing that you just said, a 30 percent cut, I think we would become more of a coastal guard.*

## Question: Could we go back to the inland waterways challenges and discuss that a bit further?

Rear Admiral Castillo: *We used to have a much more active presence inland with Boating Safety Teams. The decision was made some time ago to take that funding that supported that, give that to the states, so the states, as partners, could manage that, and the Coast Guard would focus more on the coast, still retaining some responsibilities and authorities, and primarily using the Coast Guard Auxiliary (an organization of volunteers) inland.*

*I've got a situation now where we're working with Utah. Utah's giving us I think three Utah Department of Natural Resources boats for the Auxiliary to run because we're running short on assets, and they are running short on people.*

*They've got these boats that are sitting there. It's a partnership made in heaven that they're willing to provide these boats and maintain them, so the Auxiliary can operate them. But that doesn't happen everywhere.*

*And the number of boaters on the inland ways and rivers is just phenomenal.*

**Question: The impact of national and state budgetary cuts can create a perfect storm where safety and security can be seriously compromised. What are your thoughts on the challenge of managing with missing assets?**

Rear Admiral Castillo: *There was a case here in Alameda quite recently, and I think it made national news. There was a gentleman who committed suicide by drowning in sight of the shore. He stood for some period of time in chest-deep water.*

*People on shore were watching this, and there were police and firefighters who came to the scene. Because of the budget cuts to them locally, they had discontinued their water rescue training two years prior, and nobody was trained. And there was no equipment.*

*They didn't have the requisite training and the equipment to make their own people safe to go ahead and attempt a rescue under those circumstances. The particular location that he was in, when our boat had arrived, it couldn't get anywhere near him. The water was too shallow.*

*A half-mile away was the closest that we could get.*

*And it really does help to point out that over time, many of the people that we work with have changes to their capabilities. And depending on what their particular budgetary situation is, they could reduce or drop the marine patrol; that's often one of the first things that gets cut is the marine patrol when it comes to budget shortages. We work hard at the Sector level to be aware of capability changes such as this one.*

*Cases that we would otherwise be looking for them to respond to as partners would then come to us. And I believe it is the question of what other calls are coming at the same time, whether we can pick up that load, because if you've got six calls at the same time, you can't get to all of them because you only have two assets.*

*Something's got to give.*

## Question: Next to Search and Rescue, what do you consider to be the core USCG mission today?

Rear Admiral Castillo: *My answer would be the Port Security challenge to manage and protect the Maritime Transportation System. It is the conveyor belt of the national and global economies, moving people, goods, and services around the globe, and is a core national security issue. We cannot live as a country just within our waters. There are tremendous national security implications to keeping the Maritime Transportation System running safely and efficiently.*

By Robbin Laird and Ed Gilbert
September 8, 2011

# 7

# THE VIEW FROM THE 5TH DISTRICT

IN 2011, we sat down with Rear Admiral Lee to discuss his district, which runs from New Jersey through North Carolina. We met with him shortly before Hurricane Irene hammered the East Coast, and many of the areas affected were in his district.

Rear Admiral Lee was highly visible in the media during this storm, and for those of us badly affected by it, his leadership was deeply appreciated. While media were often seen standing on the wrong side of where the storm did damage, Rear Admiral Lee and his team understood the sound-side flooding impacts of such a storm.

Rear Admiral Lee became Commander of the 5th District earlier this year. His prior operational assignments include: Commander, Deployable Operations Group (2009-2010); Commander, Sector North Carolina (2005-2007); Commander, Coast Guard Group Fort Macon (2000-2003); Commander, Group Monterey, California (1995-1997); Commanding Officer, Station Atlantic City, New Jersey (1988-1991); and Assistant Operations Officer at Group St. Petersburg, Florida (1981-1984).

His most noteworthy and personally satisfying staff assignment was as Chief, Office of Boat Forces at Coast Guard Headquarters in

Washington, DC from 2003 to 2005. His responsibilities included program management and oversight of the Coast Guard's 1,800 boats, including the development of operational doctrine, training requirements, and safety parameters. His duties also included oversight of the National Motor Lifeboat School, the only training facility in the nation specifically dedicated to heavy weather boat operations and surf training.

**Question: Could you give us a sense of the area covered by the District and some of the challenges you face?**

*Rear Admiral Lee: My AOR runs from New Jersey down through and including North Carolina. I have the mid-Atlantic region of the East Coast, including 1.5 million square miles of the Atlantic Ocean. Unlike some other Districts you have already visited, I don't contend with the significant alien migration and drug trafficking issues common to my counterparts on the southeastern and southwestern borders.* *

*That said, I do have a unique mission in the National Capital Region that none of the others have—Rotary Wing Air Intercept. In this capacity, I play a supporting role to the Northern Command (Northcom) by providing our MH-65D helicopters to intercept low and slow airborne intruders in the restricted air space around Washington, DC. In short, we are part of the layered defense in the region, and we are extremely proud to be part of the interagency team that makes this all happen.*

**Question: Could you talk about the challenges off of North Carolina?**

---

* AOR stands for Area of Responsibility.

Rear Admiral Lee: *It is called the "graveyard of the Atlantic" for good reason. The outer banks of North Carolina are notoriously treacherous, given the shoals extending miles into the sea off Cape Hatteras, Cape Lookout, and Cape Fear. There are few places to seek refuge for deep draft vessels when things go bad, particularly in the area between Cape Lookout and Cape Hatteras, leading to a considerable number of high-end search and rescue cases. Some of these cases push our people and our assets to the limits.*

*Upgrades in rescue equipment and sensors have improved our search and rescue efforts in the AOR. Our helicopters are more powerful and have better sensors (MH-65D). Our radio towers have better coverage, overlap, and redundancy and can "hear" further off the coast (R-21). Our rescue boats are bigger and more capable (47's & RBMs). The Fishing Vessel SHEILA RENEA case on January 2, 2010 is an example of our sensors, equipment, and personnel working together near Oregon Inlet to save the lives of three fishermen last winter.*

**Question: To do the missions in your District, what new capabilities would you like to have available?**

Rear Admiral Lee: *There are many needs, too many to address off the cuff. The first that comes to mind involves the mission in the National Capital Region. I need to provide my pilots in the NCR with the technology used by fighter jets to intercept aircraft that have intruded into the restricted air space. The current method of vectoring them in by radio is archaic, outdated, and I submit, dangerous.*

*As for surface vessels, we are currently looking ahead at the next generation of response boats to replace the aging fleet of 25-footers currently running nationwide.*

## Question: What about the fisheries challenges in the District?

Rear Admiral Lee: *We are an integral part of the East Coast fishing consortium. We work closely with NOAA, NMFS, and the commercial fleet to ensure, as best we can, that the regulations are appropriately enforced while not crossing the line into what some would call "excessive enforcement."*

*We are always mindful of the fact that the men and women engaged in this dangerous trade are working hard to make a living in an industry that has taken some hard hits over the years, fuel prices and dwindling stocks notwithstanding.*

*However, without Coast Guard enforcement efforts, certain species would likely be overfished to the point that the entire industry could collapse, and the American public would no longer enjoy access to plentiful seafood. It's a delicate balance.*

## Question: Could you talk about your role in tending the waterways?

Rear Admiral Lee: *We mark the nation's waterways much like the highway department marks our highways. This is what keeps commerce flowing in and out of our major ports. Without access to the ports, the economic engine of our country would soon come to a grinding halt. We exist, by and large, on trade, and more goods are carried by ship than any other mode of transportation.*

*We accomplish this mission with an aging fleet of buoy tenders, construction tenders, and small boats. It is an unglamorous, under-recognized, and under-promoted mission that we do every day, day in and day out. I simply can't overemphasize the impact that this critical mission has on the nation's commerce.*

**Question: And this is a constant activity, one requiring monitoring and resetting the buoys when, for example, storms come through?**

Rear Admiral Lee: *Yes, it is definitely not a one-off activity. When we have a hurricane come through, it does two things. It can blow your aids to navigation away, or change the channels so those aids that remain are no longer accurate.*

*We can't allow ships to start coming into port and offloading their cargo until we have verified that there is in fact a shipping lane for them to safely transit through.*

*As you know, Walmart and most of America has now shifted to a new business model that relies on "just-in-time" delivery. For every day we hold a ship offshore at anchorage waiting for the waterway to open, it costs the shipper tens of thousands of dollars. It quickly adds up as more and more ships arrive off the coast waiting for the channel to open. As the line backs up, people ashore quickly start to feel the effects.*

*For example, you have the local contractor waiting for the Italian tile that he needs to finish the house down the street. No tile, no work for his crew. You can replicate this example across the spectrum of commodities. The impact can be global when the supply chain is interrupted. That's how critical our ports are, and that's why the aids to navigation mission is so critical to the nation.*

**Question: Could you discuss some of your environmental enforcement activities?**

Rear Admiral Lee: *Part of our job is protecting people from the ocean and the ocean from the people. To that end, we do a lot of things, some of which are little known and seldom publicized. An example is the prosecution of "magic pipes."*

*I hate magic pipes. A magic pipe is an implement devised*

*by certain unscrupulous operators that is used for the specific purpose of bypassing machinery designed to separate oil and other waste in the bilges from what they pump into the ocean.*

*By using a magic pipe, the operator is seeking to bypass a functioning oily-water separator to save time and money. The magic pipe shoots the oily waste directly overboard into the ocean. We catch a number of operators doing this each year. Last year we had four or five successful prosecutions, and the fines can be in the millions. Nothing to sneeze at.*

**Comment: It sounds like you enjoy your job a lot.**

Rear Admiral Lee: *I absolutely do. We are doing vital missions very well with great people; it's a joy coming to work every day. Being on duty around the clock would be a chore in some jobs, but not this one. This is pure privilege.*

Rear Admiral Lee during our Interview.

By Robbin Laird and Ed Gilbert
September 27, 2011

# 8

# THE USCG SAN FRANCISCO SECTOR

DURING OUR WEST COAST visit in late July 2011 to examine USCG operations and challenges, we engaged in an extensive conversation with San Francisco Sector leadership. Captain Cynthia Stowe, the Sector Commander, led the discussion, joined by Captain Bliven and Commanders Stuhlreyer and Tama.

Notably, the Sector had recently inaugurated a new operations center that will significantly enhance the USCG's operational capabilities, particularly for search and rescue missions.

*U.S. Coast Guard Sector San Francisco today unveiled its advanced high-tech operations center, which officials indicated will enable multi-agency coordination during disasters and substantially improve the efficiency of search and rescue operations. The $18.1 million, two-story Interagency Operations Center (IOC), financed by the U.S. Department of Homeland Security, officially opened at a ribbon-cutting ceremony featuring keynote speaker House Minority Leader Nancy Pelosi, D-San Francisco. The Coast Guard's mission,* *she emphasized, is to protect Americans from maritime threats and the sea itself.*

*"The IOC will enable us to accomplish this mission using the most cutting-edge, innovative technology and interoperability among all first responders," Pelosi stated. "That represents a remarkable achievement." She noted that immediately following the September 11 attacks, such interoperability among local, state, and federal agencies appeared "almost impossible. Now you have achieved it here," she said.*

*The new facility includes dedicated workspace for Coast Guard personnel, U.S. Immigration and Customs Enforcement, U.S. Customs and Border Protection, and California Emergency Management Agency to work collaboratively during emergencies. Representatives from local jurisdictions including San Francisco, Alameda, and Oakland police and fire departments can coordinate operations at the interagency center, which will prove particularly valuable during the upcoming America's Cup sailing competition, according to Pelosi.*

*The new command center also houses Rescue 21, a sophisticated emergency communications and location system that Sector San Francisco has been gradually integrating into its operations, according to Coast Guard Captain Jay Jewess. This system helps Coast Guard rescue teams rapidly pinpoint distressed vessels by reducing search areas by as much as 95 percent. When mariners transmit distress calls over VHF Channel 16, the signal reaches seven radio towers positioned along the coast and inland waterways, with four additional towers scheduled to become operational soon.*

*The advanced receivers can determine the direction from which calls originate, enabling rescuers to trace direct bearing lines to distressed vessels and narrow the search area significantly, Jewess explained. The system also incorporates Channel 70 for digital communications, and many maritime radios now include embedded GPS capabilities, allowing position data to be transmitted automatically, further resolving location challenges. Previously, towers would receive signals without determining their origin point; rescuers only knew which towers had detected the signal, using that information to triangulate a much broader search area.*

*Sector San Francisco began implementing Rescue 21 approximately six months earlier, according to Coast Guard Captain Cynthia Stowe. The technology provides valuable assistance on a daily basis, Sector San Francisco personnel confirmed.** 

Captain Stowe during the interview.

**Question: What geographic area does the sector cover?**

---

* http://sfappeal.com/news/2011/06/coast-guard-unveils-new-high-tech-operations-center-in-sf.php

CAPT Stowe: *Sector San Francisco's area of responsibility extends from the northern California-Oregon border southward along the coastline for approximately 600 miles to the Monterey-San Luis Obispo County Line. Our jurisdiction extends 200 miles offshore and encompasses the entire San Francisco Bay, including the Sacramento-San Joaquin River Delta, reaching the ports of Sacramento and Stockton nearly 100 miles from the Pacific Ocean. Additionally, we maintain responsibility for inland lakes supporting interstate commerce, notably Lake Tahoe and Flaming Gorge Reservoir in Wyoming and Utah.*

**Question: This region extends deep into the continental interior while also reaching far offshore. Does this reflect the Coast Guard's mission of managing territory and addressing threats within the maritime domain?**

CDR Stuhlreyer: *Exactly. We monitor, regulate, and manage vessel movement on the water, including arrivals, departures, economic activities, and recreational activities, essentially all activities within the maritime domain.*

**Question: What assets are available for your missions?**

CAPT Stowe: *We operate four 87-foot Coastal Patrol Boats for the Sector, enabling us to maintain at least one boat underway continuously, providing single-point coverage for the AOR. We also manage eight Multi-Mission Small Boat Stations, two positioned on the Coast, two in San Francisco Bay, and two on the rivers. One station represents the Coast Guard's only alpine lake station: Station Lake Tahoe.*

*We maintain a seasonal station at Santa Cruz, California,*

*subordinate to our Monterey station. Additionally, we operate a Marine Safety Detachment in Humboldt Bay and an Aids to Navigation Team covering the Bay and inland rivers. Collectively, these resources cover one of the nation's largest geographic Sector AORs.*

## Question: We last met when you were assigned to Sector Miami. How does Sector San Francisco differ from Sector Miami?

CAPT Stowe: *The primary mission emphasis varies considerably between Miami and San Francisco AORs. In Miami, law enforcement constitutes the primary mission, including frequent drug interdiction and migrant interdiction operations.*

*For San Francisco, Search and Rescue and Port Safety and Security represent our primary focus areas. San Francisco responds to the nation's highest volume of Search and Rescue cases, exceeding 1,500 cases annually.*

*Environmental protection of the Bay and coastline also holds great importance in San Francisco. The Coast Guard works daily with other federal, state, local, and private environmental managers to eliminate or minimize threats to the marine environment.*

## Question: How does West Coast weather impact operations compared to tropical weather in Miami? What different perspective does serving here versus Miami provide regarding the physicality of the Pacific?

CAPT Stowe: *Here's an illustrative example. Last year in Miami, we had a case where two men became disoriented in an afternoon squall in their 14-foot open boat and headed away from shore until their fuel was exhausted. We located them four days later off the Georgia Coast. They were dehydrated but otherwise unharmed.*

*On the West Coast, when we lose a boater, swimmer, or kite surfer, everyone enters emergency mode, and you have a very limited time window to locate them. Additionally, there's no sandy bottom for dropping anchor and no sandy beach for landing. If you're in distress offshore, you'll end up on the rocks if we can't reach you quickly.*

CDR Stuhlreyer: *Another difference on the West Coast compared to the East is the somewhat lighter asset distribution. For instance, our small boat stations are positioned anywhere from 60 to 100 nautical miles apart along the coast, resulting in much larger AORs for each station. Along the same stretch of coastline on the East Coast, you might find two sectors and six or eight small boat stations.*

*The Pacific Coast has only a few small boat stations. This reflects historical development, geography, and population density, and creates unique challenges regarding distance and response time.*

**Question: What assets does the air station have?**

CDR Stuhlreyer: *Air Station San Francisco operates four HH-65 Helicopters—the Coast Guard's short-range rescue helicopter. For cases in the northern or southern AOR, the helicopter will likely require refueling before responding, additional refueling enroute, and can then remain on scene for a limited time. This perfectly illustrates the unique challenges posed by our extensive AOR.*

**Question: How does the water temperature here compare to Miami and how does it affect operations?**

CDR Stuhlreyer: *The water temperature is significantly colder. Offshore water temperature in the San Francisco AOR varies between 51 or 52 degrees Fahrenheit in winter, possibly reaching 57 or 58 degrees in summer. The colder temperature dramatically reduces survival time and consequently our reaction time for conducting Search and Rescue missions.*

CAPT Stowe: *Covering our geography to execute Search and Rescue missions in a timely manner presents significant challenges. Another factor stressing our resources is that in the San Francisco coastal and Bay region, we have very limited support from commercial vessel assist companies. In the Southeast, vessel assist companies like Sea Tow are highly prevalent, reducing the Coast Guard resources necessary for assisting boaters needing help.*

**Question: You need longer-range assets with enhanced speed, such as the new Sentinel Class Fast Response Cutter. What is your status for receiving these core assets?**

CAPT Stowe: *These cutters will replace the aging 110-foot patrol boats. Since San Francisco has 87-foot Coastal Patrol Boats, we aren't currently scheduled to receive a Fast Response Cutter.*

CDR Stuhlreyer: *The 87-foot patrol boats are relatively new and represent a significantly more capable platform than the previous 82-foot boats. However, in 12-14 foot seas, there's a point of diminishing returns. In such conditions, the 47-foot motor lifeboat becomes our preferred platform. With a surfman aboard, the Motor Life Boat can operate in seas up to 30 feet and 20-foot breaking surf with 50-knot winds. These are the highly capable boats frequently featured in television programs due to their rugged construction and self-righting design.*

**Question: What about helicopters and other assets?**

Commander Stuhlreyer: *The helicopters are excellent plat-forms; however, distance limitations restrict our ability to deploy the asset. We frequently depend on robust partnerships with local and state partners operating vessels and aircraft to help overcome these challenges.*

**Comment: Following 9/11, you have additional missions beyond the already demanding law enforcement and Search and Rescue responsibilities.**

CAPT Stowe: *It's certainly challenging for Multi-Mission Small Boat Stations to train, qualify, and achieve proficiency across an expanding array of missions the Coast Guard is responsible for executing. For example, consider a unit like Station Golden Gate, located just inside San Francisco Bay.*

*They're designated as a surf station, placing them at the advanced end of the training spectrum in terms of difficulty required to achieve proficiency for operating boats in the most extreme weather conditions. Station Golden Gate is also a level one PWCS unit.**

*Only one other Coast Guard Station nationwide holds both surf station and level one PWCS designations.*

**Question: What defines level one status?**

CAPT Stowe: *Level one designation means they must maintain heightened capability to respond to homeland security threats or incidents. They plan and execute homeland security missions, including patrols, escorts, and security boardings. To accomplish these missions, Station personnel train and qualify*

---

*   PWCS stands for Ports, Waterways and Coast Security.

*with additional weapons, including small boat-mounted automatic weapons.*

*Boat operators for these missions receive advanced training in boat tactics and qualification standards. Today's larger small boat stations maintain approximately the same 45-person crew as they had two decades ago when they executed only Search and Rescue and limited law enforcement. Today's mission set includes PWCS.*

**Question: So a key requirement is different training levels for the two mission sets: Search and Rescue and homeland security?**

CAPT Stowe: *You can characterize a platform as multi-mission, which is accurate; however, for PWCS operations, we require an automatic weapon mounted on the boat with a qualified operator manning the weapon, and a boat operator proficient in advanced boat tactics.*

*That configuration doesn't work effectively for assisting a disabled recreational boater adrift or approaching rocks. For that mission, we require a 47-foot MLB with a trained surfman at the helm. We genuinely must outfit for two separate missions, while the Station is staffed for a single ready boat crew.*

**Question: How do you train for the Homeland Security mission set?**

CDR Stuhlreyer: *PWCS training on the West Coast is complicated by the extensive number of marine sanctuaries and environmentally sensitive areas, essentially leaving very limited areas for this training.*

CAPT Stowe: *During the 1990s, the Coast Guard would simply proceed a short distance offshore for training. Now, due* to various regulations and policies, we transit 12 miles offshore to conduct weapons training. However, operating limitations for the 25-foot Response Boat preclude deploying it in sea conditions typically encountered 12 miles offshore in the Pacific Ocean.

Consequently, I may have a young person manning the automatic weapon who hasn't actually fired it from the 25-foot Response Boat they use for San Francisco Bay patrols. To meet training requirements, we typically train 12 miles offshore with the 87-foot Coastal Patrol Boat.

The challenge intensifies because only one area off Bodega Bay lacks sanctuaries at the 12-mile mark. All offshore training for our AOR occurs at this location, but we now understand there's a legislative proposal to expand the sanctuary to include this area, which would push Coast Guard training further offshore.

From an environmental perspective, we support protecting the environment, but we also need adequate training areas, personnel, and ammunition to support required tactical training.

CDR Stuhlreyer: We're experiencing similar challenges with HH-65 Helicopters regarding airborne use of force training. The Coast Guard equipped our helicopters with airborne use of force (AUF) capabilities.

Our aircrews who traditionally flew Search and Rescue missions now engage in special tactics training for pilots to fly the aircraft, special training to maintain gunner qualifications, and face the same challenges finding training areas.

CAPT Stowe: To help achieve proficiency for our small boat crews, we recently established a regulatory Safety Zone for Use of Force Training in San Pablo Bay. We accomplished this through a formal rule making process that included public notice and solicited public comment before publishing final regulations. This process resulted in a designated location in

*San Pablo Bay where we can conduct both airborne and waterborne use of force training. This represents a major advancement in our ability to train for the PWCS mission.*

**Question: Let's focus again on your available assets and the operational limits they impose. What is the maximum range of the 87-foot patrol boat?**

CDR Stuhlreyer: *It depends primarily on environmental conditions. Weather and sea state limit our search and rescue reach and our ability to conduct fisheries law enforcement. We conduct extensive fisheries enforcement work in conjunction with California Fish and Game and our partners at the National Marine Fisheries Service, but that cooperation tends to occur closer to the coast due to environmental conditions farther offshore.*

**Question: This significantly limits your ability to conduct security, environmental, and certainly search and rescue operations. What impact would a new Fast Response Cutter have on operational range?**
CAPT Stowe: *The FRC would provide double the endurance of the 87-foot CPB and can operate effectively out to 200 miles.*
**Comment: Let's discuss your role as Captain of the Port and how one should understand the port within the overall maritime trade system.**

CAPT Stowe: *For a port to succeed, free commerce flow is essential; goods and services must flow freely 24 hours daily, seven days weekly. Today, ports are generally limited by the depths of the channels serving them. The Port of Oakland ranks as the nation's fifth busiest container port.*
*The Army Corps of Engineers dredged the port to 50 feet several years ago, but the Corps is challenged to maintain that*

*depth within current budgetary constraints. Gradually, the controlling depths are decreasing, consequently reducing the cargo volume ships can carry.*

*Much of my role as Captain of the Port involves working with all port partners to keep the port functioning at peak capacity while vigilantly maintaining safety and security. Keeping shipping lanes clear of hazards maintains navigation routes open. When a buoy light fails, it's replaced.*

*Closing a harbor and suspending all oil transfers during a tsunami, or restricting access to an area during oil spill cleanup operations, falls under the Captain of the Port's responsibility. Likewise, determining when it's safe to reopen the port and which cargos are most critical to enter port first represent some of the many challenging decisions the Captain of the Port makes regularly.*

CDR Tama: *An important aspect of serving as Captain of the Port is serving as the facilitator who brings diverse stakeholders together to help ensure port safety, security, and efficiency. The Captain of the Port serves as a central authority where shippers, environmental advocates, local governments, and various interest groups with disparate concerns convene to resolve complex issues.*

*Without this role, there wouldn't be a single place that hears all those voices, considers all stakeholders, and makes difficult decisions about what will happen.*

**Comment: Finally, let's discuss the America's Cup coming to the Bay Area.**

CAPT Stowe: *No one consulted the Coast Guard when they decided to award America's Cup to San Francisco Bay, but as a result of that decision, the Coast Guard will establish a restricted area where sport's oldest trophy can be contested. We'll execute a plan to ensure ship movement into and out of the port during the event, host dozens of international mega-yachts, and oversee spectator safety for thousands of spectator craft. The race is expected to draw spectators second only to the Olympics.*

*It will include approximately 20 racing days in 2012 and up to 40 racing days in 2013. Tremendous planning and coordination must occur to make this happen, and that planning is seamless to the public. The Coast Guard will publish special local regulations through a notice and comment formal rule making process.*

**Question: And presumably, this is on top of everything else you're doing and complicates everything else you're doing?**

CAPT Stowe: *Correct. The Coast Guard doesn't receive any additional funding for these types of events. They're on top of our normal and emergency response operations.*

CAPT Bliven: *The largest challenge will be holding the event while keeping the Bay operating normally. Notably, the trade system needs to continue functioning normally even though the race is occurring. The container ships will want to maintain normal transit schedules, and the normal small boat users in the bay will want to operate normally. Yet a significant portion of the Bay will need to be dedicated to the race.*

CAPT Stowe: *We will also leverage this opportunity to educate boaters on proper safety procedures, since the event will attract a rather large recreational boating population. We will partner with America's Cup organizers to communicate the safety message through their website and advertisements.*

By Robbin Laird and Ed Gilbert
October 17, 2011

# 9

# THE CHALLENGE POSED BY THE NEW LARGE CONTAINER CARGO SHIPS

RECENTLY, we posted a piece looking at the impact of the new large container cargo ships on U.S. ports and facilities.* The Fabiola is only half full for this test of the port and demonstrates the challenges facing the USCG and the ports. In that piece we cited a recent LA Times article that reminded us of the emergence of the mega-cargo ships' arrival on the scene.

*The largest cargo container ship to ever dock in the Americas made a fog-shrouded first voyage into the Port of Long Beach on Friday morning, sending a message to competitors that Southern California can handle the giant vessels most others can't welcome for at least two more years.*

*Out by the breakwater, it looked as though a man-made island had sprung up overnight, but the dark shape was a vessel called the Fabiola, gliding very slowly toward port.*

*The Fabiola is one of a new generation of vessels that can*

---

* https://www.sldinfo.com/the-emergence-of-very-large-cargo-ships-another-contribution-to-"gapsmanship"/

*carry 11,000 or more containers, favored by ocean cargo lines because packing more freight boxes onto each ship lowers costs.*

*"She's way beyond our previous record for size," said Dick McKenna, executive director of the Marine Exchange of Southern California, which logs the arrival and departure of all ships calling at the ports of Los Angeles and Long Beach, the nation's largest seaport complex. "This is quite a significant jump for us."*

*The Fabiola, owned by Geneva-based Mediterranean Shipping Co., can carry 12,500 containers. The ship is just 30 feet shorter than the Empire State Building is tall, as wide as a 10-lane freeway and big enough to carry the contents of eight 1-million-square-foot warehouses.* *

We have had a chance to discuss this development with Captain Cynthia L. Stowe, Commander of U.S. Coast Guard Sector San Francisco.

**Question: Captain, how significant is the impact of the new mega container ships on the U.S. maritime transportation system?**

Captain Stowe: *It's a significant issue in terms of both logistics and waterways management because this larger class of container ship that is now arriving on the West Coast is the largest class of vessels to arrive in North America, although they're not the biggest ships that are in service right now.*

*The FABIOLA class of container ship is here as a result of even larger ships entering European trade, freeing up these ships to enter the U.S. Pacific trade. The reality is this class of ship presents a challenge in terms of the size of our existing port infrastructure and we don't have a good offshore solution*

---

* http://www.latimes.com/business/la-fi-big-ship-20120317,0,3972429,print.story

*to container delivery. These arrivals are a unique challenge to the U.S. because of the costs associated with expansion of our port infrastructure, if it's possible at all.*

*Recently, the Port of Oakland went from being the fifth busiest container port in the nation to the fourth busiest port. However, when you look at the inner and outer harbors in Oakland, the existing turning basins do not readily accommodate the size of these larger container ships.*

*The inner harbor turning basin is completely bounded by shipyard and various other facilities. We've got metal on all sides of the basin, so expansion is not possible. However, the Oakland Inner Harbor terminals are seeking larger ships due to the economical benefits that those ships bring.*

*The turning basin on the inner harbor was designed for a 980-foot ship to be able to turn. That was the original limit. About 10 years ago the harbors were deepened to 50 feet. At that time, the turning basin was reevaluated and based on enhanced navigation equipment the size of ships was increased to 1,143 feet in length. The FABIOLA measures 1,205 feet in length.*

*When the San Francisco Bar Pilots learned that the larger ships were bound for the West Coast, they conducted a simulation study at Cal Maritime Academy to determine the feasibility of bringing this larger class of vessels into the Port of Oakland. The study showed that if the pilots carried additional precision navigation equipment and the environmental conditions, i.e. winds and currents, were favorable, the larger ships could be turned in existing turning basins.*

*The equipment the pilots carry is DGPS which is temporarily mounted on the bridge wing to allow the pilot to see the precise rate of turn of the bow, stern and the pivot point of ship. If enough tug horsepower is applied and we have ideal environmental conditions, it can be done.*

**Question: Let me go back over what you have said so far to summarize. The first proposition is very simple: the ports of San Francisco and Oakland were built for a different age, pre-container, even with the anticipation of how cargo was handled kind of pre-container.**

**Secondly, obviously they've adjusted to the container age, but it's been a challenge.**

**Now you bring in a ship that is the start of a whole different class of cargo ships where you're going to have vastly increased size. The trend line is much larger ships through widened canals.**

**The trend is real and it's going to accelerate, and that means that the disconnect between the ships that are being built and will be servicing the global economy and your ports, will grow. The disconnect is a major challenge. Is that a fair way to put it?**

Captain Stowe: *That's a great summary of where we are right now. The requirements for port entry include winds less than 15 knots, current less than one knot and daylight turning only. Those conditions will challenge the speed with which we can move the ships through the Port of Oakland.*

*A number of additional conditions have to be met to bring these ships into our harbor. The pilots are requiring four tractor tugs, two 50-ton tugs, and two 80-ton tugs, which are the biggest tugs we have in the port. Additionally, the pilots determined that they need two pilots in order to safely turn the larger ships. One pilot acts as an electronic pilot utilizing the DGPS tools and advises the lead pilot on the precision navigation during the turn of the ship.*

*Additionally, impacts are being felt by other port partners when the larger container ships arrive including a requirement to stop any bunkering operations, and watch for surge interaction as the ships pass.*

*Security's an important piece of this as well because larger ships equate to the cargo being delivered at a more rapid rate. We just saw a new Supply Chain Security Strategy come out of the White House, which will help to focus our attention on the security aspects of the trade.*

## Question: How do you manage the security challenge of the container ships?

Captain Stowe: *We typically see 9,000 containers on each ship. Now, that number can be as high as 12,600 containers per ship. The Coast Guard has a very good risk-based model for determining which ships present a higher security risk. We look at the ship, where it's been, its owners, and determine which ships present a higher risk.*

*We board high risk ships at sea. Coast Guard Security Boarding Team members board these ships 12 miles offshore where they conduct crew accountability and conduct a sweep of the ship.*

*However, this doesn't account for the contents of the containers. Customs and Border Protection Agency established an international inspection system through their Container Security Initiative, which has placed U.S. inspectors around the globe to ensure the security of the contents of containers.*

## Comment: Managing the safety side of this must be very challenging as well with the limited resources at your disposal.

Captain Stowe: *One of the roles of the Coast Guard is to ensure that companies involved in shipping have established effective safety management systems for the operation of their ships while in U.S. ports. Shipping lines worldwide must make reasoned decisions about bringing their ships into our ports including an assessment of risk associated with these larger ships entering our yet to be expanded U.S. port infrastructure.*

*There will be pressure on U.S. ports to receive these larger ships and it's incumbent on all of us to make sure we don't wait for an accident to happen before we decide the ships are too large.*

*I'm currently not resourced to establish a moving safety zone around every ship coming into port. Additionally, port partners are concerned about the impact on their operations during the transit of the larger ships.*

**Question: So what are some of the resources that would be helpful to deal with this evolving challenge?**

Captain Stowe: *Above all, there is a clear need for port infrastructure investments. The demand for dredging, and port expansion has far outweighed the Army Corps of Engineers' ability to deliver. As a result, the port infrastructure hasn't kept up with demand and the larger ships are here.*

*Around the world, competent authorities tell commercial ships where they may go within their ports. However, in the U.S. we place the burden on the shipping company to assess and remain responsible for their ships' intended routes. In the future, it's clear that the larger ships will require the oversight of all involved parties to ensure the safety and environmental protection of our ports.*

**Question: You need obviously some manpower to do these tasks as well, and what kind of assets would you like to have as you go forward?**

Captain Stowe: *Post 9/11, the Coast Guard took on a greatly expanded security role in our ports. We have a certain amount of small boats and crews to execute our traditional missions and our newer security responsibilities. When we add additional responsibilities we become challenged to do everything.*

*With each of the recent arrivals of the larger container ships, the Coast Guard put a moving safety zone around the ship, we put a safety team on the ship, and assigned small boats to enforce the zone in order to keep recreational vessels at a safe distance. That equates to an increased workload.*

*There is ongoing research into the feasibility of moving the delivery of containers offshore as the ships get larger and our ports do not. The offshore element is also viewed as possibly providing enhancements in container security. If that were to occur, the Coast Guard would need assets to be able to oversee our port safety and security responsibilities.*

**Question: This would translate as well into a requirement for some new boats to manage the maritime transportation system?**

Captain Stowe: *I believe it would require a different class of vessels for us. San Francisco has four 87-foot Coastal Patrol Boats that are very light and are capable of operating in the offshore environment, but they also have a job in the ports as well. A ship with enhanced endurance would likely be necessary to continually operate 50 to 100 miles offshore. The Fast Response Cutters (FRCs) are a better look in that direction, however, none of those ships are slated to come to our AOR.*

By Robbin Laird and Ed Gilbert

May 13, 2012

# PART 3

# THE DEEPWATER
# APPROACH

Part 3 of "Always Ready, Persistently Under-Resourced" focuses on the transformation of the U.S. Coast Guard, tracing its journey through the Deepwater acquisition era, its abrupt demise, and subsequent reform efforts to adapt and modernize in the face of persistent resource constraints.

The section begins with an in-depth look at the hope and strategic vision that fueled the Deepwater Approach, highlighting its ambition to overcome block obsolescence through integrated, multidomain asset development.

Throughout, Part 3 shows how shifting policy environments, evolving threats, and inconsistent political support have complicated these modernization efforts, leaving the Coast Guard in a cycle of innovation often undermined by chronic underfunding and bureaucratic turbulence.

1

# THE DEEPWATER APPROACH

AT THE DAWN of the twenty-first century, the United States Coast Guard confronted an existential crisis that would fundamentally reshape not only its operational future but establish a a new model for defense acquisition across the American national security establishment.

Facing the simultaneous bloc obsolescence of nine classes of major assets within fifteen years, the service stood at a crossroads between accepting diminished capabilities, pursuing traditional one-for-one platform replacements, or embarking on a genuinely innovative path that would capitalize on network communications and integrated systems to create capabilities appropriate for the information age.

The USCG was way ahead of the rest of the military services in focusing on what in essence was a security not a kill web approach. I my look back at the writing on Deepwater in recent years, the fundamental perspective is missed in most analysis.

As a told a conference in Canberra a couple of years ago, the USCG shaped the first multi-domain acquisition strategy. What is

promise was not fully realized is a warning to the advocates of multi-domain acquisition efforts.

The Coast Guard chose innovation. The resulting Deepwater program represented far more than a modernization effort. It became a proving ground for capabilities-based procurement, public-private partnership, and systems-of-systems management that would offer lessons extending well beyond maritime security. This was transformation driven not by abstract theory but by operational necessity, fiscal constraint, and the recognition that traditional approaches could no longer meet twenty-first century security challenges.

What makes Deepwater particularly significant is that it was conceived and largely designed **before** September 11, 2001, fundamentally altered America's security calculus. The program emerged from careful analysis of mission requirements, technological trends, and budgetary realities rather than crisis-driven reaction. And it was crafted before the USCG was shoe horned into the new Department of Homeland Security and associated issues.

Yet when the homeland security imperative dramatically elevated the Coast Guard's role in defending the nation, Deepwater's architecture proved remarkably well-suited to the new demands. Prior to 9/11, Deepwater was about a service's innovative approach to procurement; after 9/11, it became an innovative approach to national survival.

## The Strategic Challenge: Technology Diffusion and the Erosion of Advantage

The challenge confronting the Coast Guard extended far beyond aging platforms. The service found itself engaged in a technological race against adversaries who could exploit commercial technologies for criminal and terrorist purposes more rapidly than traditional acquisition cycles could respond. The revolution in military affairs promised macro advantages to the United States in dealing with nation-state adversaries, but the diffusion of advanced technology to private groups and non-state actors created unprecedented challenges for agencies charged with homeland security responsibilities.

Drug smugglers, human traffickers, and terrorist organizations could increasingly access capabilities on the commercial market that outpaced Coast Guard operational assets. This created a troubling paradox: the nation's maritime security forces risked falling behind the very threats they were charged with confronting. Traditional Coast Guard missions, from drug interdiction and alien migrant interdiction to search and rescue and fisheries enforcement, were being outstripped by technology available to those challenging port and maritime security.

The fourteen Deepwater mission areas, spanning operations in regions generally greater than fifty miles offshore or requiring extended on-scene presence, all depended on technological superiority that could no longer be taken for granted. Providing for maritime security, dealing with drug interdiction, and executing traditional Coast Guard tasks required capabilities that could be rapidly adapted to evolving threats. The danger was clear: falling behind in the technological race with smugglers and terrorists would compromise national security at its maritime frontiers.

This challenge was compounded by severe fiscal constraints that made traditional procurement approaches financially impossible. One-for-one replacements of major platforms would have consumed resources the service simply did not possess. The Coast Guard's annual procurement budget could not support sequential replacement of aging cutters, aircraft, and systems using traditional acquisition methods. Yet declining to modernize would accept deteriorating capabilities precisely when mission demands were increasing.

## The Deepwater Approach:
## From Platforms to Capabilities

The Deepwater program fundamentally reconceptualized what platform acquisition means. Rather than focusing on specific hardware platforms, a particular class of cutter or aircraft, the Coast Guard developed performance specifications describing the fundamental capabilities needed to perform all missions in the Deepwater regions worldwide. This represented a profound shift from platform-centric to network-centric thinking.

The foundation of this approach was rigorous measurement of actual mission and operational data against requirements. The Coast Guard created a comprehensive database providing modeling and simulation capabilities to determine optimal approaches for acquiring platforms and connectors within an overall 'system-of-systems' schema.

Fourteen Coast Guard Deepwater mission areas were reviewed and validated through a Mission Analysis Report, Mission Needs Statement, and an Interagency Task Force on Roles and Missions that refined the approach and underscored the need to shift from platform-centric to network-centric operations during recapitalization.

Critically, the Coast Guard identified sixty-six measures of effectiveness spanning all Deepwater mission areas. These metrics would enable continuous evaluation of how well proposed solutions actually met mission requirements.

This performance-based approach fundamentally shifted the relationship between government and industry from one focused on delivering specified products to one focused on achieving defined outcomes. The emphasis was placed squarely on operational results rather than platform specifications.

The distinction between traditional systems integration and systems-of-systems management proved crucial. Traditional integration focuses on developing and delivering an integrated product—building an effective fighter aircraft or ship by starting with a platform and integrating systems into it.

Systems-of-systems management, by contrast, is capability-focused rather than product-focused. It asks fundamentally different questions:

- What capabilities does the client need?
- What products are available in the global marketplace to provide those capabilities?
- How might those products best be meshed to provide required capabilities now and in the future?

The driving forces for the Deepwater approach were multiple and interconnected: block obsolescence of major assets with nine existing classes reaching end of service life within fifteen years; capability limitations of current assets including lack of interoperability with the Navy and other federal agencies; increasing logistical demands due to aging assets and technology; performance measurement revealing existing capability gaps and increasing demand for Coast Guard services even before September 11th; budget realism emphasizing the need for long-term, balanced acquisition; opportunity to capitalize on new approaches to command, control, communications, computers, intelligence, surveillance, and reconnaissance (C4ISR); and opportunity to exploit relationships among system components rather than treating each as independent.

## Redefining the Government-Industry Relationship

The Deepwater acquisition strategy represented a radical departure from traditional defense procurement relationships. The Coast Guard initiated a Request for Proposals process that resulted in three competing industry teams working with the service to establish a new procurement model.

This was not the traditional dynamic of government specifying requirements and industry responding with proposals. Instead, it established a genuine public-private partnership to assess mission requirements and performance objectives in solutions offered by industry.

At the heart of this model was the concept of the lead systems integrator (LSI) working in partnership with the Coast Guard to continuously evaluate and adapt the acquisition approach. The competition itself was structured to assess not merely proposed platforms but the industry teams' approaches to defining and managing a system-of-systems solution. Proposals submitted by industry teams included comprehensive performance measurement and tracking plans integrated around their solutions.

In June 2002, the Coast Guard awarded the Deepwater contract to Integrated Coast Guard Systems (ICGS), a co-equal partnership between Northrop Grumman Corporation and Lockheed Martin Corporation. The program's total potential value over three decades could reach approximately seventeen billion dollars, making it the largest recapitalization effort in Coast Guard history. The scope was comprehensive: up to 91 ships, 35 fixed-wing aircraft, 34 helicopters, 76 unmanned surveillance aircraft, upgrades to 49 existing cutters and 93 helicopters, plus systems for communications, surveillance, and command and control.

The relationship between the Coast Guard and ICGS fundamentally differed from traditional prime contractor arrangements. The systems integrator's role oriented toward tapping into the commercial world to provide the research and development and technical refresh rates necessary for subsystems shaping the twenty-first century Coast Guard.

This open business model pursued partnerships necessary to meet Coast Guard needs rather than protecting proprietary technologies or maintaining closed development environments.

On the government side, this partnership required new capabilities and commitments. The Coast Guard had to clearly identify and communicate mission requirements while maintaining commitment and stability throughout the acquisition process.

The service retained fundamental responsibility for executing missions, assessing mission growth trajectories, and communicating closely with the lead systems integrator regarding shifting needs.

This was not delegation of responsibility but sophisticated partnership requiring active government engagement.

As Master Chief Petty Officer John Rector of the Deepwater Surface Technical Assessment Team noted, "the winning industry team and the Coast Guard were compelled to make difficult and sometimes controversial tradeoffs between upgrading legacy surface assets and achieving cost-effective design, construction, testing, and evaluation of future Deepwater surface assets."[*]

## Revolutionary Flexibility: Trading Among Platforms

The Deepwater approach sought to implement genuine capabilities-based procurement enabling continuous tradeoffs among platforms and systems. The Coast Guard and prime systems integrator team would evaluate platforms and technical solutions against mission requirements not just at program inception but continuously throughout execution. If fewer helicopters were required than fixed-wing aircraft, or if more cutters were necessary than aviation assets, or if unmanned aerial vehicles would best serve particular missions in specific areas, the program possessed the flexibility to make those tradeoffs.

This flexibility extended to the treatment of legacy systems, which the Coast Guard recognized would continue in service as new platforms were developed and fielded. Understanding that legacy assets would remain operational initially, with upgrade or replacement decisions left to the industry team in partnership with the Coast Guard, required extensive dialogue regarding status and performance baselines.

The ability to make platform tradeoffs represented acquisition flexibility that the Pentagon itself was struggling to achieve. The

---

[*] https://dspace.mit.edu/bitstream/handle/1721.1/16640/56191584-MIT.pdf?sequence=2&isAllowed=y

Coast Guard had been forced by necessity into this position: bloc obsolescence combined with fiscal constraints provided the opportunity to innovate at precisely the right moment in defense technology evolution.

The service could adapt capabilities to changing missions incrementally over thirty years, maximizing operational effectiveness while minimizing total cost of ownership. This represented a frugal but intelligent acquisition strategy allowing continuous adaptation that traditional acquisition models could not match.

## The Security Web Foundation: C4ISR Integration

At the operational heart of Deepwater lay integrated command and control and the ability to use sensor data and systems to provide a comprehensive C4ISR approach to maritime security. Integration was necessary not as a technological end in itself but as the means to enhance effectiveness of both legacy and new components working in concert. The Coast Guard would work incrementally from a new foundation to integrate systems so that operational effectiveness was maximized as operational costs were minimized.

This security CrISR web approach fundamentally distinguished Deepwater from traditional platform-based acquisition programs. A new commercial off-the-shelf (COTS) and Navy-compliant C4ISR system would net various assets together to dramatically increase maritime domain awareness. This 'system-of-systems' architecture enabled information sharing and coordinated operations that individual platforms could never achieve in isolation. Data fusion would be critical, as would interoperability with other U.S. agencies and allied forces.

The emphasis on commercial technology was strategically essential, not merely cost-driven. The pace of commercial marketplace innovation meant that military-specific development often resulted in fielding equipment less capable than adversaries could purchase in the open marketplace. By defaulting to COTS and commercially available non-developmental items (CANDI), Deepwater could tap into rapid commercial technology refresh cycles while maintaining interoperability with Navy systems and other federal agencies.

The C4ISR architecture provided the foundation for continuously leveraging new commercial technologies throughout the program's lifecycle. The efficient use of resources, ability to tap into changes in the commercial marketplace, capacity to plug into a mix of mission sets, and provision for much lower lifetime ownership costs with high technological refresh rates represented significant advantages of the Deepwater approach.

The cooperation extended to information technology initiatives with the U.S. Navy. The Coast Guard's Deepwater program was designed to work closely with the Navy's IT-21 strategy and network warfare initiatives. The Joint Navy/Coast Guard Policy Statement of February 2001 envisioned building a National Fleet of multi-mission surface combatants, major cutters, patrol boats, and aircraft to maximize effectiveness across all naval and maritime missions. This interoperability was critical to national and global missions, enabling seamless operations whether Coast Guard forces operated independently or as part of joint task forces.[*]

## Strategic Depth: Pressing Out Our Borders

The Deepwater program embodied a strategic concept articulated as 'Pressing Out Our Borders', dealing with threats to the homeland and maritime security as far from United States shores as possible. This vision envisaged a layered defense comprising surveillance, detection, identification, sorting, interception, and engagement of threats in four

---

[*] https://www.globalsecurity.org/military/library/report/2002/mil-02-06-wave lengths03.htm;   https://www.usni.org/magazines/proceedings/2002/august/deepwa ter-will-provide-homeland-security

areas of approach to the United States: overseas source departure zones, trans-oceanic route zones, U.S. coastal route zones, and U.S. port zones. By operating at greater distances from American shores, threats could be thwarted well before reaching positions to deliver attacks against the nation.

This strategic vision required close planning liaison and operational coordination with the Navy and emphasized the Coast Guard's role at the nexus of domestic and international security responsibilities. The challenge to the Coast Guard could only increase in decades ahead from technological revolution and its diffusion to private groups. Meeting this challenge demanded capabilities extending well beyond traditional maritime law enforcement to encompass comprehensive maritime domain awareness and the ability to project security operations far from U.S. shores.

The new cutters, surveillance assets, and C4ISR systems central to Deepwater would allow the Coast Guard to execute this strategic concept much more effectively than legacy systems permitted. The National Security Cutter, vertical unmanned aerial vehicles, and Maritime Patrol Aircraft would provide capabilities for sustained operations in distant waters while maintaining connectivity with shore-based command centers and other operational forces. This combination of platform capability and network connectivity represented genuine transformation in Coast Guard operational reach.

## September 11th: Validation and Urgency

The tragic events of September 11, 2001, fundamentally transformed the context for Deepwater implementation while validating the wisdom of the approach. The program had been conceived and designed before those events to meet traditional Coast Guard missions more effectively and cost-effectively. The common misunderstanding that Deepwater was created in response to terrorism obscures an important reality: the Coast Guard had pursued innovative modernization driven by operational necessity and fiscal constraint, not by any specific threat scenario. Deepwater was about meeting traditional missions better; 9/11 made it about national survival.

September 11th dramatically elevated the Coast Guard's significance in homeland defense and validated the Deepwater approach's design. The demands on the service to provide maritime security against terrorists became the number one task, making the need to extend homeland security defense urgent.

President Bush captured this urgency in June 2002: "The Coast Guard's Deepwater Program will award a multi-year contract to replace aging ships and aircraft, and improve communications and information sharing. The whole purpose is to push out our maritime borders, giving us more time to identify threats and more time to respond."[*]

The post-9/11 operational tempo created severe stresses revealing both the urgency of modernization and the wisdom of the Deepwater approach. Resource hours for ports, waterways, and coastal security surged from a pre-9/11 baseline of 19,291 hours to 254,640 hours by fiscal year 2003. a more than thirteen-fold increase.

Simultaneously, the age of existing assets created dangerous operational conditions. The Coast Guard's fiscal year 2003 safety review revealed an in-flight engine power loss rate of 62.74 per 100,000 flight hours, far exceeding both FAA definitions of probable event occurrences and Navy Safety Center guidelines.[†]

These pressures led to recognition that accelerating Deepwater funding would enable moving new assets into position more rapidly while reducing reliance on dangerous legacy systems. The value-for-money approach inherent in Deepwater would not be modified by acceleration; rather, additional funding would enable faster transition

---

[*]   https://www.usni.org/magazines/proceedings/2002/august/deepwater-will-provide-homeland-security
[†]   https://www.gao.gov/assets/a241785.html

to new capabilities while reducing high logistics and operations costs associated with aging legacy assets.

The system's strategic fundamentals, common C4ISR providing interconnectivity and leveraging commercial technologies, enabled rapid capability delivery impossible through dedicated military specifications approaches.

In short, the promise of the Deepwater program represented far more than Coast Guard modernization. How the Coast Guard operated in the present and would need to operate in the future formed the baseline for considering modernization options.

Congress had demanded that the Pentagon and military services take a 'big picture' look at meeting homeland and national security requirements. Deepwater offered a vision of what defense transformation could achieve.

The Coast Guard chose a revolutionary path when confronting bloc obsolescence, and in doing so, created a model for how defense organizations might navigate the perpetual challenge of maintaining operational superiority in an era of accelerating technological change and evolving security threats.

But it eventually failed to deliver on its promise. Such multi-mission thinking is the order of the day now in defense circles. But Deepwater was the first multi-domain approach but it failed to achieve its core vision of how to transform an integrated force.

The next chapter addresses this issue.

# 2

# THE DEMISE OF THE DEEPWATER APPROACH

WHILE THE PROGRAM'S vision was revolutionary, its execution revealed fundamental tensions between ambition and capability, between innovation and institutional readiness, and between the demands of transformational acquisition and the organizational realities of a service in transition.

The scope of what the Coast Guard attempted with Deepwater cannot be overstated. This was not simply an effort to replace aging ships or upgrade aircraft; it was a holistic reimagining of how the service would operate in the twenty-first century. The program sought to create an integrated system of systems, where new cutters, helicopters, fixed-wing aircraft, unmanned aerial vehicles, communications networks, and command centers would work seamlessly together to extend the Coast Guard's operational reach and effectiveness.

The integration of C4ISR capabilities represented a quantum leap in sophistication for a service that had traditionally operated with far more modest technological infrastructure.

What made Deepwater particularly bold was that no other American military service had attempted such comprehensive modernization on this scale all at once. The Navy, Air Force, and Army typically pursued platform-specific acquisition programs, replacing ship classes or aircraft types in sequential, manageable increments.

The Coast Guard, by contrast, was attempting to replace virtually its entire deepwater fleet and aviation assets while simultaneously implementing new technologies and operational concepts. This all-encompassing approach reflected both visionary thinking and, as events would prove, considerable risk.

## The Acquisition Capability Gap

To understand why Deepwater encountered such significant difficulties, one must first appreciate the Coast Guard's limited experience with complex, large-scale acquisition programs prior to the program's launch. The service's traditional procurement approach bore little resemblance to the sophisticated acquisition management practiced by the Department of Defense's military branches.

Historically, the Coast Guard acquired platforms in small quantities, often on an irregular basis. When it did need major vessels or aircraft, it frequently piggybacked on Navy procurement programs, essentially purchasing variants of systems the Navy had already developed and tested.

This approach, while cost-effective and low-risk, meant the Coast Guard had never developed the institutional expertise required to manage a massive, integrated acquisition program. The service lacked a deep bench of experienced acquisition professionals. Its contracting infrastructure was modest.

Its program management methodologies were not designed for the complexity of a multi-domain, multi-platform effort extending across decades. The Coast Guard simply had never needed to orchestrate holistic upgrades across multiple operational domains simultaneously, and thus had never built the organizational muscle memory for doing so.

Deepwater demanded capabilities the Coast Guard was still developing. Managing such a program required a highly skilled acquisition workforce capable of understanding complex technical requirements, negotiating sophisticated contracts, managing multiple prime contractors and subcontractors, integrating disparate systems, conducting rigorous testing and evaluation, and maintaining program control amid inevitable changes and challenges. These were not skills the Coast Guard possessed in depth at the program's outset.

Compounding these challenges was the program's adoption of a "lead systems integrator" approach. Recognizing its own limitations, the Coast Guard contracted with private industry, specifically a joint venture between Lockheed Martin and Northrop Grumman—to serve as the LSI, effectively outsourcing much of the program's design, integration, and even some source selection decisions.

While this approach promised to leverage private sector expertise and efficiency, it created a fundamental problem: the Coast Guard was ceding control over major aspects of its own modernization to contractors.

The LSI model reduced the Coast Guard's institutional learning. When contractors make key decisions about requirements, design, and procurement, government program managers gain less hands-on experience and develop less institutional knowledge.

Moreover, the inherent conflict of interest in having a contractor both design the overall system architecture and select vendors for individual components created concerns about objectivity and cost control. The LSI had incentives to design requirements favoring its own capabilities or those of favored partners, potentially driving up costs while limiting the government's insight into alternatives.

## The DHS Integration Challenge

Just as Deepwater was ramping up in the early 2000s, the Coast Guard underwent a massive organizational upheaval that would profoundly affect the program's trajectory. Following the September

11, 2001 terrorist attacks, Congress created the Department of Homeland Security, consolidating twenty-two previously separate federal agencies into a single department focused on protecting the homeland. In 2003, the Coast Guard was transferred from the Department of Transportation, where it had resided since 1967, into this new mega-department.

This organizational realignment, while perhaps logical from a homeland security perspective, created significant challenges for Deepwater. The Department of Homeland Security lacked expertise in overseeing multi-billion-dollar, decades-long defense acquisition programs.

Unlike the Department of Defense, which had evolved sophisticated acquisition oversight structures, standardized program management methodologies, and deep institutional knowledge of military procurement, DHS was essentially starting from scratch.

The department's senior leadership came primarily from law enforcement, immigration, emergency management, and other civilian backgrounds. Few had experience with the unique challenges of acquiring complex military systems.

The acquisition processes and resources at DHS were not tailored for military system procurement. The department's focus was understandably on its core homeland security missions: border security, immigration enforcement, disaster response, cybersecurity, and counterterrorism. Major defense acquisition programs like Deepwater represented a small fraction of DHS's overall portfolio and did not receive the specialized attention they required.

Where the Department of Defense had entire organizations dedicated to acquisition oversight, with robust testing and evaluation infrastructures and sophisticated cost estimating capabilities, DHS was building these capabilities on the fly while simultaneously standing up an entire department.

Funding became increasingly uncertain as the Coast Guard competed for resources within a department whose priorities lay elsewhere. The predictable, multi-year funding streams essential for major defense acquisitions were difficult to maintain when the Coast Guard was one voice among many agencies vying for DHS dollars. Budget uncertainties led to program stretches, inefficient procurement strategies, and difficulty maintaining critical workforce skills as experienced personnel left due to program instability.

Moreover, DHS priorities increasingly shifted Coast Guard attention toward homeland security missions and away from the defense-specific, long-term planning that Deepwater's scale demanded.

The Coast Guard found itself pulled in multiple directions, supporting port security, enforcing immigration laws, conducting drug interdiction, responding to hurricanes, and maintaining its traditional maritime safety missions, while simultaneously trying to execute a transformational modernization program. The service's resources and leadership attention were stretched thin.

## Post-9/11 Mission Expansion

The September 11 attacks not only triggered the Coast Guard's move to DHS but also fundamentally expanded the service's mission requirements. Suddenly, the Coast Guard found itself on the front lines of homeland defense, with new responsibilities for port security, maritime domain awareness, and preventing terrorist attacks via the maritime domain. These new post-9/11 mission requirements led to expansions in Deepwater's scope during execution, further straining an already stressed program.

What had been conceived as a recapitalization effort now had to incorporate enhanced capabilities for detecting and deterring terrorist threats, improved interoperability with other homeland security and defense agencies, and technologies for tracking and identifying vessels in congested coastal waters.

These requirements added complexity to system designs, increased costs, and extended development timelines. The program was essentially being replanned on the fly, even as acquisition contracts were being executed and early assets were entering production.

This mission expansion fostered even greater reliance on contractors, as the Coast Guard's own acquisition staff struggled to keep pace with evolving requirements. Cost overruns became endemic. Schedule slips mounted.

High-profile problems emerged, including the notorious failure of the 123-foot patrol boat conversion program, where structural problems forced the Coast Guard to decommission newly refurbished vessels. The National Security Cutter program, Deepwater's flagship effort, experienced significant delays and cost growth.

Congressional oversight intensified as Deepwater's troubles became public. Government accountability reports documented management failures, inadequate oversight, and questionable contractor performance. The program's reputation suffered, and "Deepwater" became synonymous in some circles with acquisition dysfunction rather than innovation.

## Shaping a Transitional Way Forward

The turning point came when the Coast Guard began making fundamental changes to how Deepwater was managed. Recognizing that the LSI model was not working, the service restructured the program, essentially terminating the integrated LSI approach and breaking Deepwater into separate platform-specific acquisition programs. Each major asset class, the National Security Cutter, Offshore Patrol Cutter, Fast Response Cutter, maritime patrol aircraft, and helicopters, became its own program with dedicated government program managers exercising direct control.

Critically, the Coast Guard began integrating experienced DoD acquisition professionals into key leadership positions. These individuals brought decades of experience managing major defense acquisition programs and understood the methodologies, best practices, and pitfalls of complex procurement. They implemented more rigorous requirements development processes, strengthened technical oversight, improved cost estimating, and established clearer lines of accountability.

The service also invested heavily in building its own acquisition workforce. Training programs were established, career paths were clarified, and acquisition became recognized as a core competency requiring sustained investment. The Coast Guard Acquisition Directorate was professionalized, with clearer standards, better tools, and stronger support from service leadership.

As these changes took effect, program performance improved markedly. The National Security Cutter program stabilized and began delivering capable ships. The Fast Response Cutter program became a model of efficient acquisition, delivering vessels on time and on budget. New aircraft programs proceeded more smoothly.

While Deepwater would never fully escape its troubled reputation, the latter phases demonstrated that with proper management, organizational commitment, and adequate resources, the Coast Guard could successfully execute major acquisition programs.

## The Impact

Despite its difficulties, Deepwater produced significant long-term value. The program forced the Coast Guard to modernize its acquisition culture and develop capabilities essential for any service procuring complex military systems.

The organizational learning that occurred, often painfully, built a foundation for future success. Today's Coast Guard acquisition workforce, while still developing, is far more capable than the organization that launched Deepwater at the turn of the century.

Many assets ultimately procured under or evolved from Deepwater have proven vital to Coast Guard operations. The National Security Cutters represent a quantum leap in capability over the vessels they replaced, with enhanced endurance, better seakeeping, sophisticated sensors, and improved aviation facilities. The Fast

Response Cutters have become workhorse patrol vessels, replacing aging 110-foot patrol boats with modern, more capable platforms. New aircraft and command centers have enhanced the Coast Guard's operational effectiveness across its diverse missions.

The program initiated a crucial shift to systems thinking within the Coast Guard. Even if the original vision of a fully integrated system of systems was not completely realized, the service learned to think more holistically about how different platforms, sensors, and networks need to work together. This systems perspective influences how the Coast Guard approaches modernization today, even in more incremental programs.

The lessons learned from Deepwater have influenced not just the Coast Guard but the broader acquisition community. The program's experience demonstrated the risks of the lead systems integrator model and the importance of maintaining government program management expertise.

It highlighted the challenges of transformational acquisition and the dangers of expanding requirements during execution. It showed how organizational context—including which department a service resides in—can profoundly affect acquisition outcomes.

## Conclusion

The Deepwater program's reputation reflects both its innovative vision and the severe systemic challenges it faced. The ambition to comprehensively modernize an entire service through a single integrated program was breathtaking in scope.

The concept of creating a security web spanning multiple domains anticipated the joint all-domain operations concepts that would later become central to American military thinking. The Coast Guard's leadership showed courage in attempting such a transformation.

However, ambition alone cannot overcome the realities of acquisition immaturity and challenging organizational positioning. The

Coast Guard simply was not ready, in terms of workforce, infrastructure, or experience, to manage a program of Deepwater's complexity.

The service's placement within a newly formed department focused primarily on civilian homeland security missions, rather than within the defense establishment, created additional obstacles. The timing, coming just as post-9/11 mission requirements were exploding, added further strain.

The program's evolution from troubled mega-program to restructured portfolio of more manageable efforts reflects a maturation process both in the Coast Guard's acquisition capabilities and in broader understanding of how to approach large-scale military modernization.

Today's more incremental, platform-focused approach may lack the visionary sweep of original Deepwater, but it reflects hard-won wisdom about matching ambition to capability.

The Deepwater experience reminds us that in acquisition, as in maritime operations, respecting the environment you operate in is crucial for success. The Coast Guard's deepwater operating environment demanded modernization, but the acquisition environment the service found itself in during the 2000s with limited institutional capacity, uncertain organizational positioning, and expanding mission requirements was far more challenging than the open ocean.

Navigating those troubled waters required the Coast Guard to learn painful lessons, make difficult course corrections, and ultimately chart a more sustainable path forward. The service that emerged from the Deepwater experience is stronger, more capable, and better prepared for the acquisition challenges of the twenty-first century, even if the journey was far rougher than originally anticipated.

# 3

# ADMIRAL JOHN CURRIER
# AND ACQUISITION REFORM

THE YEAR 2007 stands as a watershed moment in United States Coast Guard history, a year when the service faced its most severe acquisition crisis and responded with fundamental institutional reform.

At the center of this transformation stood Rear Admiral John P. Currier, whose technical expertise, operational experience, and strategic vision helped shepherd the Coast Guard from the troubled Deepwater program to a disciplined acquisition enterprise. His role in this transition, though not always widely recognized, proved instrumental in reshaping how the service would acquire and support its assets for decades to come.

The Deepwater program, launched in the late 1990s, represented the Coast Guard's ambitious attempt to address what it called "block asset obsolescence" or the simultaneous aging of cutters, aircraft, and systems across its entire fleet. Rather than pursuing traditional acquisition approaches, the service opted for an integrated "system-of-systems" strategy, partnering with a commercial lead systems integrator to manage the entire recapitalization effort.

By 2007, this approach had encountered severe problems. The

management and execution of what was then a single, integrated Deepwater program faced strong criticism from various observers. House and Senate committees held oversight hearings examining program failures. The conversion of 110-foot patrol boats proved unsuccessful. Cost overruns mounted. Schedule delays accumulated. The fundamental structure of the program with its reliance on a commercial systems integrator and fragmented Coast Guard oversight showed critical weaknesses.

Admiral Thad Allen, who became Commandant in May 2006, recognized that the service needed more than incremental fixes. He issued directives to transform the acquisition system fundamentally, calling for the consolidation of acquisition organizations and the integration of logistics and support systems within a new directorate. This set the stage for what the Coast Guard would call its "Blueprint for Acquisition," released in February 2007.

John Currier brought exceptional credentials to the acquisition reform challenge. Promoted to flag rank in 2005, he was designated Assistant Commandant for Acquisition at Coast Guard Headquarters precisely as the Deepwater crisis intensified. His background uniquely prepared him for this role.

Currier had been designated an Aeronautical Engineer in 1982 and served as Engineering Officer at Coast Guard Air Stations in Traverse City, Michigan and Astoria, Oregon. Critically, he had served as Deputy Program Manager (Engineering) for the Coast Guard HH-60J and Navy HH-60H joint helicopter acquisition at the Naval Air Systems Command in Washington, D.C. This experience gave him intimate knowledge of how the Navy approached acquisition, knowledge that would prove invaluable in reshaping Coast Guard practices.

But Currier was no mere technocrat. With over 6,000 flight hours in Coast Guard and Navy aircraft, he had commanded Air Station Detroit and Air Station Miami, the world's busiest air-sea rescue unit. He understood operations viscerally. He had served as Pacific Area Chief of Operations and Area Chief of Staff. He knew what operators needed and how acquisition failures affected the men and women executing missions. This combination, technical depth, acquisition experience, and operational credibility, gave him the authority to drive fundamental change.

In my April-May 2007 interview with Admiral Currier for Military Logistics International, he outlined the principles guiding the Coast Guard's acquisition transformation. These principles, codified in the Blueprint for Acquisition, would fundamentally reshape how the service approached asset recapitalization.

Four core principles drove the reform effort.

- First, program managers and integrated project teams (IPTs) would become the central units of action in acquisition. Rather than fragmented responsibility spread across multiple organizations, clear accountability would rest with empowered program managers.
- Second, the Coast Guard would significantly enhance both the numbers and capabilities of its acquisition workforce, uniformed and civilian alike. The service recognized that effective acquisition required professional expertise that it had allowed to atrophy.
- Third, acquisition would become a more effective enabler of operations. Requirements generation, definition, and execution needed to evolve into a true partnership between acquisition and operations, rather than the disconnected processes that had contributed to Deepwater's problems.
- Fourth, and Currier emphasized this was arguably most important, acquisition would be reshaped from the perspective of life-cycle support. The Deepwater approach had attempted to contain logistics within the program itself. The new approach would integrate acquisition and sustainment within a unified mission support architecture.

As Currier explained, this represented "a major change in thinking for the Coast Guard." Where Deepwater had created a separate acquisition silo, the new approach would align acquisition with the broader mission support structure. A three-star officer would lead this mission support organization, placing it on par with the operations directorate, itself a radical move for Coast Guard culture.

On July 13, 2007, the Coast Guard established its consolidated Acquisition Directorate (CG-9). This organizational change, which Currier helped design and initially led as a flag officer, brought together previously fragmented acquisition entities under unified leadership and common processes.

The new directorate consolidated the portfolio of major and minor acquisition projects, contracting and procurement functions, research and development programs, logistics support and transition to sustainment functions, and other acquisition support elements under a single command. More importantly, it established an acquisition governance structure with strengthened processes and began building a highly capable and trained acquisition workforce.

The launch represented more than bureaucratic reshuffling. As Currier emphasized in our 2007 discussion, the new structure placed program managers at the center of the acquisition process. The Research and Development Center was incorporated into an Office of Research, Development, Test and Evaluation, specifically designed to support program managers rather than operate independently. Throughout, life-cycle support became the organizing principle.

Currier stressed that implementation would take time, three years to achieve real integration of acquisition and support. He acknowledged the cultural challenges: "this is not the time to throw out the baby with the bathwater." The Coast Guard needed to maintain its ability to acquire capacity while transforming how it did so. Operational demands continued; the service couldn't pause missions while restructuring itself.

One of Currier's most important contributions was his clear-eyed assessment of the Coast Guard's limitations and requirements. As a mid-sized service, the Coast Guard would have limited ability to perform systems integration itself. But it needed the capability to govern the process and work effectively with government partners or commercial systems integrators.

This represented a fundamental shift from the Deepwater model. Rather than outsourcing systems integration responsibility to a commercial partner, the Coast Guard would retain governance capability while leveraging external expertise. The service would empower program managers and contract officers to operate in common but evolving processes, creating what Currier described as a "permanent revolution" focused on delivering timely, supportable capability.

The new structure would enable three different contractual approaches: acquisition of goods and services, with sustainment integrated into the asset acquisition process; single asset acquisition, managed by empowered program managers; and systems integration, where the Coast Guard would govern rather than execute the integration work.

The establishment of the Acquisition Directorate in 2007 marked the beginning, not the end, of Coast Guard acquisition transformation. Currier's subsequent career tracked the evolution of this reform effort and demonstrated its effectiveness.

After his tenure as Assistant Commandant for Acquisition, Currier commanded Coast Guard District 13 in the Pacific Northwest. In 2009, he became Chief of Staff of the Coast Guard, later transitioning to become the service's first Deputy Commandant for Mission Support, the three-star position he had described in our 2007 interview as essential to putting acquisition and logistics on par with operations.

In this role, Currier continued to oversee acquisition reform. By 2011, when I interviewed him about the National Security Cutter program, the transformation he had helped launch was bearing fruit. The NSC program, which had experienced early challenges, had matured into what Currier called "a cost-controlled and risk managed program." Hull #4 was contracted at a fixed price. Hull #5 was coming in at virtually the same cost. The program demonstrated stable requirements, controlled costs, and a contractor with a positive learning curve, precisely what the acquisition reforms were designed to achieve.

As Currier explained in that 2011 interview, the NSC represented more than just a ship. It was a deployed security system, a floating command post capable of hosting task forces, deploying small boats and helicopters, processing intelligence, and managing threats ranging from narcotics interdiction to disaster response. The successful acquisition and deployment of this complex capability demonstrated how far the Coast Guard had come from the Deepwater crisis.

When Currier became Vice Commandant in May 2012, serving until his retirement in May 2014, he continued as the Component Acquisition Executive, providing strategic oversight of the acquisition enterprise he had helped create. In congressional testimony during this period, he could point to concrete achievements: the Coast Guard had delivered operational capability on-cost, on or ahead of schedule, and in a controlled risk environment. Coast Guard acquisition personnel received four 2012 DHS Acquisition Awards, including Acquisition Professional of the Year and Component Acquisition Executive Team of the Year.

At the heart of the reform effort was enhanced capacity for integrated project management within an evolving mission support architecture. As Currier emphasized in our 2007 conversation, the existing situation featured requirements definition, project management, and contracting processes that were too disjointed and disconnected. By empowering program managers and rebalancing relationships between contract managers and program officers, a more effective process could emerge.

The key outcome would be better through-life management of systems and equipment, leading to lower operating costs and higher availability. This represented a fundamental philosophical shift—from viewing acquisition as simply buying assets to understanding it as establishing sustainable capability.

The new structure established fully integrated processes allowing program managers to focus, coordinate, and strategically manage projects. The mission support architecture enabled acquisition to interact strategically over an asset's life to more efficiently produce required capability. Requirements would be reviewed regularly to revalidate or redefine them with realistic judgments about affordability and technical feasibility.

Integrated Project Teams (IPTs) would exist throughout a system's lifecycle, with core competencies remaining in place even as leadership changed as systems matured. This ensured continuity and institutional memory, precisely what the fragmented pre-2007 structure had lacked.

The Coast Guard acquisition transformation that Currier helped lead offers several enduring lessons.

- First, effective acquisition reform requires leaders with both technical depth and operational credibility. Currier's background as an aeronautical engineer, his experience with Navy acquisition processes, and his extensive operational flying career gave him the authority to drive change across cultural boundaries.
- Second, organizational structure matters, but only if accompanied by process reform and workforce development. The creation of the Acquisition Directorate provided necessary unification, but the emphasis on program manager empowerment, integrated project teams, and life-cycle management transformed how the organization functioned.
- Third, mission support, including logistics and sustainment, must be integrated with acquisition from the beginning rather than treated as an afterthought. The

Deepwater program's attempt to manage logistics separately proved unsustainable. The new approach made life-cycle support an organizing principle.

- Fourth, patience and persistence are essential. Currier emphasized in 2007 that full implementation would take three years. In reality, the transformation continued throughout his career and beyond. Fundamental institutional change cannot be rushed, particularly when operational demands continue unabated.

- Fifth, effective acquisition requires clear governance of systems integration rather than wholesale outsourcing of responsibility. The Coast Guard's recognition that it needed to govern integration processes rather than perform all integration work internally reflected realistic assessment of capabilities and requirements.

I had the privilege of working with Admiral Currier over many years, from that initial 2007 interview through his time as Deputy Commandant for Mission Support and Vice Commandant. He possessed a rare combination of technical mastery, strategic vision, and genuine humility. He never sought credit for the transformation he helped lead, always directing attention to the acquisition workforce and to operational commanders.

Currier understood that acquisition existed to serve operators, to provide the men and women executing missions with capable, supportable assets. His thousands of hours flying search and rescue missions, his experience commanding air stations, his intimate knowledge of what worked and what failed in the field, these informed every decision he made as an acquisition leader.

He passed away on March 1, 2020, at his home in Traverse City, Michigan, at age 68. The Coast Guard he helped transform continues to benefit from the architecture he helped build. The professional acquisition workforce, the empowered program managers, the integration of acquisition and mission support, the emphasis on life-cycle management, these endure as his legacy.

In short, the 2007 Deepwater crisis could have crippled Coast Guard recapitalization for a generation. Instead, it catalyzed fundamental reform that positioned the service for successful acquisition in the 21st century. Admiral John Currier stands as a key figure in this transformation, though his role has not received the recognition it deserves.

From his position as Assistant Commandant for Acquisition during the critical 2007 period, through his subsequent leadership as Deputy Commandant for Mission Support and Vice Commandant, Currier helped shepherd the Coast Guard from crisis to capability. The Blueprint for Acquisition, the establishment of the Acquisition Directorate, the emphasis on integrated project management and life-cycle support, these represented not merely organizational changes but fundamental shifts in how the Coast Guard understood and executed acquisition.

The National Security Cutters, Fast Response Cutters, and other platforms successfully delivered under the reformed acquisition system stand as testament to this transformation.

But perhaps the most important legacy is the cultural change: the recognition that effective acquisition requires professional expertise, clear governance, empowered program managers, and integration of operations, acquisition, and sustainment within a unified mission support architecture.

Admiral Currier's technical expertise, operational credibility, and strategic vision proved essential to this transformation. The Coast Guard's ability to execute its diverse and demanding missions today rests in no small part on the acquisition foundation he helped build during the service's most challenging period. That represents a legacy worthy of recognition and remembrance.

PART 4

# THE LEGEND CLASS CUTTER AND THE CHALLENGES FACING U.S. SHIPBUILDING

Part four of the book presents the Legend-class National Security Cutter (NSC) as the centerpiece of Coast Guard modernization and uses its story to illuminate the wider dysfunctions and missed opportunities in U.S. naval shipbuilding.

The section opens by arguing that the NSC is not just a hull replacement for the Hamilton-class high-endurance cutters but a qualitatively different system built around C4ISR, endurance, and command-and-control, enabling the ship to function as a "chaos management" platform rather than a single-mission combatant.

Against this operational success, the book highlights the Legend class and the Navy's missing frigate as shortfalls of the U.S. shipbuilding and acquisition system. Admiral John Currier's role is central in setting a good course in motion: Part IV links his post-Deepwater reforms, creation of a professional acquisition directorate, life-cycle management focus, empowered program managers, to the stabilization of the NSC program after early turbulence. By 2011–2013, the NSC is held up as proof that disciplined requirements, realistic technologies, and coherent governance can deliver complex ships on a stable cost curve, in contrast with the earlier

Deepwater crisis. At the same time, the section underlines how the Coast Guard's chronic under-resourcing and political "penny-pinching" narrowed the class to ten hulls, limiting the ability to exploit the design as a true fleet backbone.

The most provocative strand in Part IV is the retrospective on the Navy's frigate saga. The author argues that by the mid-2000s the Navy could have adopted a militarized NSC as its next frigate, gaining an affordable, proven, blue-water hull suited to escort, presence, and low-to-medium intensity combat, but instead pursued the Littoral Combat Ship and then a drawn-out FFG(X) process. The Navy's eventual decision to base its FFX frigate on the NSC design is a "coming home" to the option that was available two decades earlier, at the cost of lost time, strategic position, and industrial stability that can never be fully recovered.

Throughout Part IV, the Legend-class story is used to knit together several themes: the Coast Guard's quiet innovation in multi-domain, networked operations; the importance of ship size, endurance, and C2 for modern maritime security; the fragility of U.S. naval shipbuilding policy; and the centrality of time over money in defense preparedness.

The section closes by arguing that the NSC program demonstrates what is possible when operational logic, acquisition discipline, and industrial capability are aligned and, conversely, how rare that alignment has been in recent American shipbuilding history.

# 1

# THE NATIONAL SECURITY CUTTER: 2010 AND 2011

BETWEEN 2010 AND 2011, as the first National Security Cutters entered operational service, a profound transformation in Coast Guard capabilities was becoming evident. Through extensive interviews conducted aboard the USCGC Bertholf and Waesche, with commanding officers, executive officers, and senior Coast Guard leadership, a picture emerged of a vessel that represented far more than a simple replacement for aging High Endurance Cutters. The National Security Cutter was revealing itself as a floating command post, a chaos management system, and a multi-mission platform designed to meet the uncertain threats of the twenty-first century.

## Beyond Hull Replacement: A Systems Revolution

The most critical insight from these early interviews was the imperative to think about the NSC not merely as a new hull, but as an integrated command and control system. Admiral Currier, then Deputy Commandant for Mission Support, emphasized this distinction repeatedly. The NSC represented an entry into the world of C4ISR enablement, where the bridge functioned as a "cockpit" for command decision-making in ways that the Hamilton-class cutters could never achieve.

Captain Lance Bardo, commanding officer of the USCGC Waesche, went further, describing the NSC as a "chaos management system." This characterization captured something essential about the vessel's purpose. In a world of unpredictable threats, from narco-terrorism to natural disasters, from mass migration to weapons of mass destruction incidents, the United States needed maritime assets capable of rapid response, sustained operations, and integrated command across multiple agencies and partners.

The bridge of the Bertholf provided a vivid demonstration of this capability. Captain John Prince described a midnight interdiction off the coast of Panama where his ship, despite having a broken helicopter, successfully pursued and intercepted a high-speed suspect vessel using radar tracking, Forward-Looking Infrared systems, and vectored small boats. The entire operation unfolded on the bridge displays, with Prince able to monitor both his pursuing small boat and the suspect vessel simultaneously, maintain navigational awareness to avoid hazards, and communicate with Combat Information Center, tactical operations centers, and national assets by pressing buttons on his command console.

"Even though it is the middle of the night, I can see what my boat is doing. I can see what the other boat is doing. I can actually watch and record what is happening," Prince explained. This level of situational awareness, impossible aboard previous cutters, transformed the NSC into what might best be understood as a mobile task force command center.

## The Operational Bubble:
## Command at Sea

Commander David Ramassini, executive officer of the Bertholf, articulated the NSC's capabilities in terms that highlighted its unique position across the spectrum of maritime operations. The ship could travel over 1,500 nautical miles in two days and remain on station for twenty to twenty-five days without refueling concerns. More importantly, it brought core C4ISR systems that could interoperate seamlessly with Department of Defense and Homeland Security partners.

The concept of the NSC as an "operational bubble" emerged from these discussions. Admiral Currier described it as a mobile security force with 360-degree awareness, able to sprint to threats or loiter for extended periods with the sensors and communications to protect large swaths of ocean or coastline. Within this bubble, the NSC commander could access domain awareness in air and surface dimensions, deploy tools to affect the outer edges of the operational area, utilize organic intelligence capability, and import critical intelligence from national systems.

During Exercise Northern Edge off Alaska, the Bertholf demonstrated this capability by simultaneously tracking over fifty aircraft at ranges exceeding two hundred miles using its three-dimensional air search radar. This kind of awareness, combined with the ability to communicate globally while managing local tactical operations, gave the NSC a flexibility unprecedented in Coast Guard history.

Captain Bardo emphasized that the NSC was large enough and capable enough to actually manage complex response operations. "We have nothing else in the Coast Guard or the Navy for that matter to manage domestic response the way this platform can," he noted. The Navy, while a capable partner, did not have domestic response as its primary mission. The NSC filled a unique niche as an asset that could operate as the centerpiece of a unified command structure, whether responding to an earthquake in San Francisco, an airliner crash off Cape Cod, or a weapons of mass destruction attack in New York Harbor.

## Multi-Mission Heritage
## and Adaptive Design

Admiral Currier provided crucial historical context for understanding the NSC's design philosophy. The Coast Guard's multimission cutter tradition stretched back to the 1930s Secretary-class 327-foot cutters. These vessels, originally built for security and interdiction, had adapted to serve as ocean station platforms, providing navigation fixes and weather information for primitive transoceanic airliners. When World War II erupted, these same cutters were weaponized and became the most effective anti-submarine warfare assets the United States possessed in the early war years.

The Hamilton-class 378-foot cutters followed this tradition, serving in Vietnam, through the Cold War submarine threat, during the marijuana and cocaine smuggling wars, and in countless rescue operations. The rescue of over five hundred souls from the liner Prinsendam off Alaska stood as one example of their contribution over forty-five years of service. But the Hamiltons had never been designed as floating command posts for combined security and defense operations.

The NSC represented the next evolution in this adaptive tradition. Built for an era when threats ranged from narco-terrorism to mass migration, from maritime terrorism to natural disasters, the NSC embodied flexibility through enhanced capabilities rather than through weapons fit-outs. As Captain Bardo explained, having endurance, stability, command capabilities, and flexibility inherent in the platform created synergistic interactions with the various tool sets the ship could deploy.

## Speed, Endurance, and
## Operational Economics

The operational advantages of the NSC became clear through detailed discussions of its propulsion system and endurance characteristics. Captain Prince described a ship equipped with two diesel engines and one gas turbine, offering five different modes of propulsion. In harbor mode, a single diesel engine drove both shafts at slow speed. Cruising mode used one diesel at higher speed for sixteen to eighteen knots. The ship could achieve approximately twenty-four knots on diesels alone, twenty-six knots on the gas turbine, and about thirty knots with combined diesel and gas turbine operation.

This propulsion flexibility translated into operational advantages that were difficult to overstate. When a tactical commander requested the Bertholf arrive on station a thousand miles away sooner than planned, the ship easily made over twenty knots using only diesel engines, arriving early with plenty of fuel for mission execution. For a displacement hull, hull speed equals 1.4 times the square root of the ship's length in feet. Any speed above that requires exponential horsepower increases. By making the ship larger, the NSC achieved more speed, more efficiently and economically than smaller cutters could manage.

Commander Ramassini highlighted how endurance enhanced operational security. By operating comfortably for nearly thirty days and loitering for forty-plus days with fuel reserves, the NSC reduced port calls and kept adversaries uncertain about its location. Every port call compromised operational security in the information age. The ability to maintain persistent presence without revealing position gave the NSC a significant advantage in counter-narcotics operations where smugglers actively sought to avoid Coast Guard assets.

The efficiency gains extended to routine operations as well. Captain Prince described fisheries enforcement operations off Hawaii where the NSC completed three boardings in a day, covering vast areas of the Exclusive Economic Zone at fifteen knots without overtaxing the propulsion plant or consuming significant fuel. A 378 or 270 would have completed only one or two boardings in the same period without using the gas turbine at three times the fuel consumption.

## Aviation Operations
## and Platform Stability

The NSC's aviation capabilities represented another quantum leap. Captain Bardo noted that the flight deck was literally twice as large as older cutters, four thousand square feet versus twenty-three hundred on a 378. The NSC could land all varieties of helicopters that the 378 could not, including the H-60. More importantly, the ship's stability allowed helicopter and unmanned aerial vehicle operations in sea states that would ground older cutters.

The aviation fuel capacity told its own story. Where the 378 carried eight thousand gallons, the NSC carried thirty-five thousand gallons, allowing extended air operations without fuel management concerns. The ship featured two hangars capable of housing two helicopters while maintaining a clear deck for landing a third. This flexibility extended to joint operations; the Bertholf trained with Army Special Forces preparing for overseas deployment, demonstrating the ship's ability to operate with high-end Department of Defense partners.

Captain Prince emphasized that aviation and small boat operations defined the NSC's purpose. The ship provided a stable launch and recovery platform for the helicopters and boats that actually interdicted smugglers, rescued persons in distress, enforced law, and ensured security. During the Bertholf's Eastern Pacific deployments, the ship was never out of pitch and roll limits for helicopter operations and never beyond Captain Prince's or his executive officer's comfort limits for small boat operations. This consistent mission-execution capability kept the ship in the "end game" business continuously.

## Small Boat Operations
## and End Game Capability

The NSC's small boat complement represented another dimension of its expanded capability. The ship carried three Over-the-Horizon small boats compared to two on legacy cutters, providing both redundancy and additional operational capacity. Captain Prince described the tactical doctrine developed for go-fast interdictions, which always involved launching two boats simultaneously. The pursuit boat served as the primary chase vessel, its sole focus on stopping the target vessel and maintaining positive control over it. The high-speed chase subjected crews to severe physical pounding as boats bounced through waves at speeds exceeding thirty-five knots.

Once the pursuit boat stopped the suspect vessel, the support boat arrived with fresh boarding teams who had not endured the battering of the high-speed chase. These personnel, rested and alert, took over the tactical situation, conducting the actual boarding and evidence collection while pursuit boat crews recovered from their physically demanding role. This division of labor recognized the reality that extended high-speed operations degraded crew effectiveness.

Commander Ramassini emphasized that while the NSC typically employed only two boats simultaneously, the third boat provided crucial backup for casualties or complex scenarios demanding three boats in the water simultaneously. This redundancy proved essential during operations involving multiple suspect vessels or when extended interdiction operations required boat rotation to maintain crew effectiveness.

The NSC's superior seakeeping qualities directly enhanced small boat operations in ways that statistics alone could not capture. Captain Prince emphasized that during Eastern Pacific deployments, sea conditions never exceeded his or his executive officer's comfort limits for launching or recovering small boats.

This consistent capability kept the ship perpetually in mission execution mode, able to conduct interdiction operations regardless of sea state. A helicopter might be down for maintenance, but the small boats remained available. Multiple air assets might deploy aboard, but small boats provided the assured capability. This redundancy across air and surface dimensions meant the NSC rarely faced circumstances where it could detect threats but lacked means to prosecute them.

The operational tempo of small boat operations aboard the NSC demonstrated the platform's human factors advantages. Crews could maintain sustained boarding operations over extended periods without the debilitating fatigue that plagued older cutters.

The combination of improved crew rest facilities, more stable launch and recovery platforms, reduced transit times to operating areas, and overall quality-of-life improvements meant that boarding teams remained sharp and effective throughout ninety-day patrols. This human engineering, often overlooked in discussions focused on sensors and propulsion systems, proved as critical as technological capabilities in determining operational effectiveness over extended deployments.

## Combat Systems Integration
## and Stealth Characteristics

The integration of sensors and systems aboard the NSC created capabilities that transformed tactical operations. The Forward-Looking Infrared system proved invaluable, allowing over-the-horizon tracking of targets of interest without revealing the cutter's position. From high on the mast, operators could visually assess targets day or night without the target detecting the NSC, even when radar tracks provided only nebulous returns that could not be correlated to specific vessels.

Commander Ramassini highlighted the NSC's relatively low radar cross-section. Older cutters presented large radar signatures, clearly visible to their targets. The NSC's design made it appear as a fishing vessel on radar at distance. This stealth characteristic, coupled with FLIR capability, allowed operations in a much more covert manner before deploying boarding teams for close engagement.

The Combat Information Center provided access to secure networks and intelligence feeds unavailable on the bridge. Captain Prince noted that while he could conduct operations from the bridge, CIC offered SIPR access and other tools that enhanced the command team's ability to integrate intelligence from various sources.

This reach-back capability proved critical during the pursuit of a fully submersible drug smuggling submarine. By communicating across different levels of government and accessing various intelligence communities, the Bertholf maintained proximity to the threat until the smugglers scuttled their submarine.

## Proven Return on Investment

The operational results from the first deployments validated the NSC's design and capabilities. Captain Prince reported that in a single ninety-day Eastern Pacific patrol, the Bertholf conducted seven interdictions, disrupting an estimated 12,500 kilos of cocaine with a street value approaching half a billion dollars. The smugglers employed a full spectrum of vessels: single-engine boats close to shore and further offshore, multi-engine vessels in various zones, fishing vessels, semi-submersibles, and fully submersible submarines. The Bertholf successfully engaged threats across this entire spectrum during a single patrol.

Admiral Currier noted that the combined interdictions had already disrupted drug shipments with street values equivalent to the cost of the National Security Cutter, profits that would otherwise have enriched cartels. This represented a remarkable return on investment in the early years of NSC operations.

## Crew Efficiency and Habitability

The NSC achieved these capabilities with striking crew efficiency. Commander Ramassini emphasized that despite being forty feet longer than the 378-foot cutters, the NSC operated with fifty fewer personnel. The crew complement represented only about ten percent more than the 270-foot Medium Endurance Cutters, yet the NSC delivered capabilities far exceeding the combined capacity of either previous class.

Captain Bardo highlighted the human factors engineering that made extended deployments sustainable. Crew fatigue had always limited endurance in older cutters. The NSC featured a dedicated ballast system that maintained ship stability as fuel was consumed, providing a more comfortable ride and reducing crew fatigue. Berthing arrangements placed personnel in staterooms with four or five roommates rather than the crowded compartments of older ships. After three weeks at sea, crews remained effective rather than exhausted.

These quality-of-life improvements had operational significance. A fresh, alert crew maintained higher levels of performance throughout extended patrols. The ability to make water and electricity self-sufficiently, combined with comfortable living conditions, allowed the NSC to remain on station for up to ninety days, functioning as a small city and global command platform.

## Crisis Management Across Scenarios

The versatility of the NSC emerged most clearly when examining its application across different crisis scenarios. Admiral Currier described how the operational bubble concept applied differently depending on the threat or emergency. In a mass migration scenario, the NSC deployed to the Florida Straits as a command and control platform capable of achieving air and maritime domain awareness with communications to exercise multi-unit command over diverse assets.

For a disaster response like an earthquake in San Francisco, the NSC could sail under the Golden Gate Bridge and establish a protected, self-sustained command node capable of co-joining unified commands with Department of Defense joint task forces. The ship's chemical, biological, and nuclear protection systems allowed operations in contaminated environments, providing responders with sustained and protected communication facilities.

Captain Bardo emphasized the Coast Guard's demonstrated excellence in domestic emergency response, citing recent examples from the Deepwater Horizon oil spill and the Haitian earthquake. The NSC amplified this capability, bringing command resources that neither the Coast Guard nor the Navy possessed for managing domestic crises. While the Navy provided capable partnership, domestic response was not its primary mission. The NSC filled a unique role as the maritime asset optimized for this requirement.

## Strategic Context and Global Security

The strategic value of the NSC extended beyond specific missions to questions of national sovereignty and access. Admiral Currier described the requirement for persistent presence off both coasts, in the Bering Sea, the Eastern Pacific, and the Caribbean for interdiction of migrants and narco-terrorism while ensuring safety and security at sea. The formula had traditionally required twelve High Endurance Cutters at 185 days per year away from homeport, based on over forty years of 378-foot cutter operations.

Through crew rotation concepts, the Coast Guard modeled achieving 230 days per year at sea with eight NSCs, roughly matching the days engaged in mission execution that twelve Hamiltons provided.

But this comparison understated the true capability increase. The NSC could loiter or sprint, carry multiple aircraft, deploy armed interdiction boats, and eventually operate unmanned aerial vehicles. It processed intelligence as a deployed system rather than simply providing days at sea from a hull.

Commander Ramassini emphasized the NSC's unique contribution to combatant commanders globally. Foreign nations worked with the Coast Guard in its law enforcement capacity, allowing operations at the constabulary end of the spectrum while maintaining the ability to interoperate with U.S. assets up to higher intensity operations. This provided access that purely military vessels could not achieve, serving as a unique instrument for regional stability and global security.

The Bering Sea operations illustrated this strategic dimension. The NSC could cover a 300,000 square mile area, steaming to the Maritime Boundary Line and throughout the Aleutian Chain, projecting U.S. presence and protecting exclusive economic zones in harsh conditions.

Commander Ramassini noted the importance of showing peer competitors that the United States maintained concern and presence in the Arctic along the Russia border. The NSC's remarkable helicopter launch and recovery capabilities in the challenging Bering Sea environment made this presence credible and sustainable.

## Acquisition Challenges and Future Trajectory

Despite operational success, the NSC program faced significant acquisition challenges. Admiral Currier reported that after early program structure and process difficulties, the acquisition had matured into a cost-controlled and risk-managed program. Hull Four had been contracted for a fixed price of $480 million unit cost, with a total acquisition cost of $692 million including long-lead material, post-production certifications, personnel costs, training, and associated expenses. The Coast Guard was about to award the contract for Hull Five at or very close to the cost of Hull Four.

The program had achieved a fixed-price environment with stable requirements and a contractor showing positive learning curves. Regular contract awards maintained production efficiency.

However, Admiral Currier warned of the real possibility that funding disruptions could force significant cost increases for hulls six, seven, and eight. Work stoppages would result in loss of yard expertise, workforce layoffs, and the cascade of negative effects that gaps in production always created in shipbuilding.

The Coast Guard also sought regulatory relief regarding OMB Circular A-11, which required total contract funding in the year contracts were signed. Shipbuilding required multiple years to complete efficiently, with long-lead material in one year, production contracts the next, and post-production activities in the third year.

This reality caused first-year money to expire before expenditure. The Navy received relief from A-11 interpretation for carrier, submarine, and frigate acquisition through either appropriations or authorization language. The Coast Guard sought parity in policy application to achieve the procurement efficiencies that Navy models yielded.

## A New Paradigm for Maritime Security

The interviews from 2010 and 2011 captured the National Security Cutter at a pivotal moment. The early operational deployments had proven the ship's capabilities across a spectrum of missions from counter-narcotics to fisheries enforcement, from military exercises to preparation for disaster response.

The platform had demonstrated return on investment through drug interdictions, operational efficiency through crew rotation and endurance, and strategic value through persistent presence in vital maritime zones.

More profoundly, these early years revealed that the NSC represented a new paradigm for thinking about maritime security assets.

Captain Bardo's description of the ship as a "chaos management system" captured its essence better than any technical specification.

In an uncertain world facing unpredictable threats, the United States needed maritime platforms capable of rapidly deploying to crises, establishing command and control over diverse assets, maintaining operations for extended periods, and adapting to whatever mission requirements emerged.

The Coast Guard's multi-mission cutter tradition, stretching back to the 1930s and proven through world war, cold war, and peacetime emergencies, had evolved to meet twenty-first century requirements. As Commander Ramassini succinctly concluded, "The Coast Guard has something special here... She's got legs and a crew who knows how to use them."

The National Security Cutter, as revealed through these interviews from its first years of operation, represented more than the sum of its systems, sensors, and propulsion plants.

It embodied a vision of maritime security for an era of persistent threats, rapid response requirements, and the need to project sovereignty across the global commons and approaches to the United States. The early operational record suggested that this vision was being realized in steel, systems, and the skilled crews who took these remarkable vessels to sea.

"THE NEW USCG CUTTER: A "Chaos Management System", August 8, 2010. This was the interview with Captain Bardo.[*]

"The National Security Cutter Opportunity," August 2, 2011. This was the Currier interview.[†]

"On the Bridge of the Bertholf: Con-Ops Enablement by the

---

[*]  https://sldinfo.com/2010/08/the-new-uscg-cutter-a-key-tool-for-crisis-management/

[†]  https://sldinfo.com/2011/08/the-national-security-cutter-opportunity/

NSC," August 8, 2011. This was the interview during a visit to Coast Guard Island in Alameda with Captain John F. Prince and Executive Officer Commander David W. Ramassini.*

"Aboard the USCG Bertholf: Operations and Capabilities of the National Security Cutter," September 13, 2011. This provides additional information from the interview during a visit to Coast Guard Island in Alameda with Captain John F. Prince and Executive Officer Commander David W. Ramassini.†

---

* https://sldinfo.com/2011/09/aboard-the-uscg-bertholf-operations-and-capabilities-of-the-national-security-cutter/

† https://sldinfo.com/2011/09/aboard-the-uscg-bertholf-operations-and-capabilities-of-the-national-security-cutter/

# 2

# THE NATIONAL SECURITY CUTTER: FROM ACQUISITION TURBULENCE TO OPERATIONAL SUCCESS (2011-2025)

IN AUGUST 2011, Vice Admiral John Currier testified before Congress about the National Security Cutter program's remarkable transformation. After years of Deepwater acquisition chaos, the NSC had achieved cost stability and predictable production rhythms. The fourth, fifth, and sixth ships were contracting at consistent costs around $480-692 million per hull.

The program appeared to have overcome its troubled origins and entered a mature phase that would deliver exactly what the Coast Guard needed: a fleet of eight highly capable cutters to replace twelve aging Hamilton-class vessels while maintaining equivalent operational presence through enhanced capabilities and crew rotation concepts.

Yet Admiral Currier also warned of the threat that would ultimately constrain the program: funding instability. "The problem is the real possibility that our funding might be disrupted," he cautioned. "This would mean that as we seek to acquire hulls #6, 7, and 8, if we are exposed to funding delays, we're going to incur significant cost increases, loss of expertise in the yard due to gapped work stoppages, people laid off, all of those negative effects."[*]

Fourteen years later, in June 2025, those warnings proved prophetic when the Coast Guard and Huntington Ingalls Industries announced they were scrapping the eleventh NSC over contract disputes.

The NSC program ended with ten delivered hulls rather than the eleven that had been funded or the eight originally planned. This chapter examines the path from Admiral Currier's 2011 testimony to the program's 2025 closure, analyzing how acquisition stability, operational demands, strategic opportunities, and funding realities shaped the NSC's ultimate trajectory.

## Stabilization and Production Success (2011-2017)

The period from 2011 to 2017 represented the NSC program's most successful phase. Building on the Coast Guard's 2007 decision to terminate the Integrated Coast Guard Systems lead systems integrator arrangement and assume direct acquisition management, the service achieved genuine program stability. The shift from performance-based acquisition to detailed specifications, combined with building internal acquisition expertise, had wrung cost and schedule uncertainty out of the process.

The results were tangible. USCGC Stratton (WMSL-752) was delivered in September 2011 and commissioned in March 2012, having been christened by First Lady Michelle Obama. Hamilton (WMSL-753) followed in September 2014, with Joshua James (WMSL-754) delivered in June 2015. Munro (WMSL-755) came in December 2016, and Kimball (WMSL-756) in February 2018.[†] This

---

[*]  https://www.dhs.gov/news/2013/12/11/written-testimony-us-coast-guard-house-transportation-and-infrastructure

[†]  https://www.naval-technology.com/projects/legendclassnsc/

production cadence demonstrated the stable contractor relationships and learning curves that Admiral Currier had described.

Technical challenges emerged but were addressed systematically. Structural analysis of the first two hulls revealed that some components could be expected to survive only three years under the demanding North Pacific and North Atlantic operational profile, where Coast Guard cutters typically operate 180-230 days annually in some of the Northern Hemisphere's roughest seas.

Structural reinforcements were incorporated in Stratton and subsequent hulls, though the third ship still suffered corrosion and leaks within weeks of commissioning in 2012._These were typical challenges for a new ship class, manageable through continuous improvement rather than fundamental redesign.

## Operational Excellence
## and Mission Validation

As the NSC fleet grew, operational tempo demonstrated the ships' value across their intended mission spectrum. Counter-narcotics operations in the Eastern Pacific became the program's most visible success metric. The NSCs' combination of helicopter support, fast interceptor boats, 60-day endurance, and command facilities made them ideal platforms for disrupting transnational criminal organizations.

By 2025, NSC operations in the Eastern Pacific had achieved historic results. The Trump administration's Operation Pacific Viper, launched in August 2025, leveraged NSC capabilities to accelerate interdiction operations. USCGC Munro conducted the largest single at-sea cocaine interdiction in 18 years, seizing over 20,000 pounds.[*] USCGC Stone offloaded approximately 49,010 pounds—the most cocaine seized by a single cutter in one patrol in Coast Guard histo-

---

[*] https://www.cbsnews.com/news/coast-guard-seizes-cocaine-single-boat-record-pacific-ocean/

ry.[*] USCGC James interdicted over 46,500 pounds valued at nearly $350 million during a 92-day deployment.[†]

Through Operation Pacific Viper, the Coast Guard seized over 150,000 pounds of cocaine in just five months, averaging more than 1,600 pounds interdicted daily.[‡] Fiscal year 2025 total cocaine seizures reached approximately 510,000 pounds, surpassing any previous year on record.[§] These operations vindicated Admiral Currier's 2011 vision of the NSC as a force multiplier capable of organizing task forces with helicopters, small boats, and international partners to achieve effects no single platform could deliver alone.

## The Strategic Opportunity Not Seized: Indo-Pacific Operations

While counter-narcotics operations demonstrated NSC capabilities, a parallel strategic opportunity went largely unexploited. Admiral Currier's 2011 testimony had explicitly identified Western Pacific operations as one of four primary NSC operating areas, noting the ten-day transit from Alameda allowed maximizing time on station in theater. Yet the Obama administration's contemporaneous "pivot to Asia" never translated into sustained NSC presence in the region.

The strategic logic was compelling. China's expanding maritime assertiveness, particularly in the South China Sea and Pacific island nations' exclusive economic zones, demanded response mechanisms beyond gray-hulled Navy warships. The Coast Guard's white hulls carry none of the threatening symbolism of naval vessels, making NSCs ideal for engagement with Pacific island nations, fisheries

---

[*]   https://www.globalsecurity.org/security/library/news/2025/11/sec-251119-uscgo1.htm

[†]   https://www.dvidshub.net/news/553696/coast-guard-cutter-james-conducts-counter-drug-patrol-eastern-pacific-ocean

[‡]   https://www.news.uscg.mil/Press-Releases/Article/4355243/coast-guard-seizes-150000-pounds-of-cocaine-through-operation-pacific-viper-int/

[§]   https://www.newsweek.com/record-breaking-cocaine-seizure-made-in-eastern-pacific-11168239

enforcement partnerships, and sustained presence operations. The service's law enforcement authorities enable capacity-building activities with partner nations' coast guards, exactly the below-threshold engagement that builds influence in great power competition.

By 2023, recognition of this gap prompted action. Vice Admiral Andrew Tiongson, Pacific Area Commander, announced that after a single NSC deployment to the region in 2022, the Coast Guard was tripling deployments to three national security cutters in 2023. Hawaii-based USCGC Kimball deployed to operate with the Japan Coast Guard under U.S. 7th Fleet tactical control. Tiongson also announced plans to deploy a 270-foot medium endurance cutter to the Pacific permanently for persistent presence operations with island nations.[*]

Yet this expansion came a decade after the pivot to Asia was announced, and remained constrained by fleet size limitations. The Coast Guard's 2011 fleet mix analysis concluded that even the planned 91 cutters (eight NSCs, 25 Offshore Patrol Cutters, 58 Fast Response Cutters) would provide only 61% of the capability needed to fully perform statutory missions.[†]

Adding Indo-Pacific presence missions atop existing counter-narcotics, fisheries, migration, and defense support responsibilities meant zero-sum tradeoffs. The gap between strategic requirements and available resources that Admiral Currier had warned about in 2011 had become acute by the 2020s.

## Funding Instability and Cost Growth (2013-2024)

The stable production rhythm achieved by 2013 proved fragile. Admiral Currier's concern about OMB Circular A-11 requirements

---

[*]   https://news.usni.org/2023/02/15/coast-guard-moving-cutter-to-pacific-as-regional-missions-expand

[†]   https://www.congress.gov/crs_external_products/R/HTML/R42567.web.html

that total contract funding be available in the year a contract is signed created structural impediments. Shipbuilding inherently requires multi-year funding for long-lead materials, production contracts, and post-production activities. While the Navy routinely received A-11 relief through appropriations language allowing incremental funding, the Coast Guard lost such relief, making efficient acquisition "very difficult."

The consequences manifested in cost growth. The 2014 Coast Guard estimate for eight ships reached $5.474 billion, averaging $684 million per ship, up from the $692 million total acquisition cost Currier cited for hull four. The sixth NSC jumped to $735 million. By 2025, Congressional Research Service estimates placed average NSC procurement cost at approximately $670 million per ship.

Congress ultimately funded eleven NSCs rather than eight, with the ninth, tenth, and eleventh ships not requested by the Coast Guard. On its face, this suggested Congress prioritized the capability.

But the additions came without commensurate increases to overall Coast Guard acquisition budgets, creating zero-sum competition with the Offshore Patrol Cutter and Fast Response Cutter programs. The FY2018 appropriations bill included $1.24 billion for NSC work covering the tenth ship, long-lead materials for the eleventh, and eleventh ship construction, spreading resources thin across multiple hulls simultaneously.

Admiral Currier's 2011 warning about funding gaps eroding the positive learning curve proved prescient. Breaks in consistent funding forced Huntington Ingalls to manage production inefficiently, laying off specialized workers and disrupting supplier relationships. When production resumed, costs increased as the yard rebuilt capabilities.

The stable, repeatable processes achieved by 2013 never translated into the economies of scale and continuous improvement that decades of consistent production had delivered for Navy programs like the Arleigh Burke-class destroyers.

## The End: USCGC Friedman
## Cancellation (2025)

The NSC program's conclusion came in June 2025 through a contract dispute that encapsulated the program's broader challenges. The Coast Guard and Huntington Ingalls Industries had contracted for both the tenth ship (Calhoun) and eleventh ship (Friedman) in a $930 million contract option in December 2018. Calhoun delivered in 2023 and commissioned in April 2024. But Friedman, which began fabrication in May 2021 with expected 2024 delivery, stalled amid what HII described as "contract-related disputes" over "material conformance concerns."[*]

By March 2025, construction had halted with the ship approximately 15% complete. Rather than resolve the technical and contractual issues, both parties agreed to cancel construction. "We worked collaboratively with the Coast Guard to reach a mutually acceptable resolution that supports and aligns with the Coast Guard's overall cost-saving objectives," HII spokeswoman Kimberly Aguillard stated. "In mutual agreement with the USCG, we have signed a contract modification that identifies an alternate strategy related to the sunsetting of the NSC program, which has already exceeded the original acquisition objective of eight ships."[†]

The settlement returned $260 million to the government while providing the Coast Guard $135 million in spare parts to maintain the existing ten-ship fleet.[‡] The FY2024 budget request confirmed program closure, seeking only $7 million for post-delivery activities and close-out support rather than funds for a twelfth ship. Though the Department of Homeland Security Appropriations Act 2020 had

---

[*] https://news.usni.org/2025/06/05/ingalls-coast-guard-scrap-11th-national-security-cutter-over-contract-impasse-says-hii

[†] https://news.usni.org/2025/06/05/ingalls-coast-guard-scrap-11th-national-security-cutter-over-contract-impasse-says-hii

[‡] https://seawaves.com/construction-of-uscgc-friedman-cancelled/

made $100.5 million available for twelfth-ship long-lead materials, this option was not pursued.

## The Current NSC Fleet: Distribution and Missions

Despite the turbulent acquisition path, the ten delivered NSCs constitute the Coast Guard's most capable surface fleet. The ships are strategically distributed across three homeports reflecting primary operational areas.

Four NSCs, Calhoun, Hamilton, James, and Stonem are stationed in Charleston, South Carolina, primarily supporting Atlantic and Caribbean operations including counter-narcotics and migration interdiction. Four NSCs, Bertholf, Waesche, Stratton, and Munro, are based in Alameda, California, conducting Eastern Pacific counter-narcotics operations and Western Pacific deployments. Two NSCs. Kimball and Midgett, homeport in Honolulu, Hawaii, positioned for both Eastern Pacific and Indo-Pacific missions.*

Each NSC features advanced command, control, communications, computers, intelligence, surveillance and reconnaissance equipment; aviation support facilities capable of operating helicopters and unmanned aerial vehicles in high sea states; a stern cutter boat launch for rigid-hulled inflatable boats; and 60-day endurance for extended station keeping. The ships' Sensitive Compartmented Information Facilities enable classified intelligence processing, making them genuine command posts for combined operations as Admiral Currier envisioned.

---

* https://www.dcms.uscg.mil/Portals/10/CG-9/Acquisition%20PDFs/Factsheets/NSC_0625.pdf

## Lessons Learned and
## Implications for Future Programs

The NSC saga from 2011 to 2025 offers several critical lessons for defense acquisition.

- First, Admiral Currier's 2011 warnings about funding stability proved entirely accurate. The program achieved genuine acquisition discipline by 2013, with stable costs and predictable schedules. But sustained stability requires sustained funding commitment. Inconsistent funding profiles and zero-sum budget competitions eroded the learning curves and supplier relationships that make serial production efficient. The positive trajectory from 2011-2017 demonstrated what was possible; the subsequent trajectory demonstrated how fragile those gains were without institutional commitment to maintaining production momentum.
- Second, strategic rhetoric without commensurate resources produces missed opportunities. The pivot to Asia represented sound strategy for countering China's expanding influence in the Indo-Pacific. The Coast Guard's unique capabilities, white-hull diplomacy, law enforcement authorities, capacity-building expertise, were ideally suited to the sustained, below-threshold engagement the strategy required. NSCs could have provided persistent presence in Pacific island exclusive economic zones, fisheries enforcement partnerships, and combined exercises with regional coast guards throughout the critical 2011-2020 decade when Chinese influence was expanding most rapidly. Instead, the service struggled to maintain minimal presence across all statutory missions, only achieving modest Indo-Pacific expansion by 2023 at the cost of accepting shortfalls

elsewhere. The gap between what Admiral Currier articulated as possible in 2011 and what resource constraints allowed in practice represents a strategic opportunity cost.

- Third, the Coast Guard's position spanning Department of Homeland Security and Department of Defense creates structural vulnerabilities in budget processes. Operating under DHS in peacetime but with Title 10 authorities for Defense Department operations, the service lacks both Pentagon advocacy and clear homeland security budget priority. Its missions extend from drug interdiction to Arctic operations to naval warfare support, making it essential across multiple domains yet chronically underfunded for its actual mission scope. The NSC program existed at this uncomfortable intersection—delivering capabilities valuable for both DHS and DOD missions yet receiving neither the Navy's budget certainty nor focused homeland security investment.

- Fourth, acquisition reform is achievable but requires sustained leadership commitment. The Coast Guard's 2007-2013 transformation from Deepwater chaos to NSC stability demonstrated that troubled programs can be salvaged through management discipline, internal capability-building, and rigorous process control. Building the Acquisition Directorate, drawing on Navy expertise, shifting to detailed specifications, and maintaining close contractor oversight produced tangible results. But these improvements required continuous attention from service leadership and consistent funding from Congress—both difficult to maintain across multiple administrations and budget cycles.

- Fifth, the temptation to add ships beyond the program of record creates false economy. Congress funding eleven

NSCs rather than eight appeared to strengthen the program. In reality, it spread acquisition dollars across more hulls without increasing total budgets, creating competition with other critical programs and preventing the service from building a truly robust fleet of any single platform type. The theoretical advantage of more platforms was offset by practical disadvantages: incomplete coverage of mission requirements across all cutter types, disrupted production rhythms, and ultimately the cancellation of the eleventh ship anyway. A more coherent approach would have matched procurement decisions to realistic total resources rather than making incremental additions that looked generous but created systemic problems.

## Conclusion: What Was Achieved and What Was Lost

The National Security Cutter program delivered ten highly capable warships that substantially enhanced Coast Guard operational capacity. These vessels serve as Admiral Currier envisioned: floating command posts for combined operations, platforms for sustained presence in critical maritime regions, and force multipliers that organize task forces exceeding the capability of any single ship. Their counter-narcotics achievements alone, 510,000 pounds of cocaine seized in FY2025, multiple historic interdiction records, vindicate the original operational concept and investment.

Yet the path from Admiral Currier's 2011 testimony to the 2025 Friedman cancellation reflects profound institutional failures. His warning about funding instability came to pass. His vision of Western Pacific operations received only belated, limited implementation. The acquisition stability achieved by 2013 never translated into the decades of consistent production that build truly efficient industrial base.

The strategic opportunity to leverage Coast Guard capabilities during the pivot to Asia was largely missed. And the program ended not with the planned eight ships, not with the funded eleven ships, but with ten ships and a cancelled hull representing both wasted resources and lost capability.

As the Navy now potentially looks to NSC-derived designs for its post-Constellation frigate program, the irony is sharp. The Coast Guard achieved what the Navy's Constellation program failed to deliver: a mature, proven design with nearly two decades of operational experience and ten delivered hulls demonstrating production capability.

Yet that very production line has been closed, workforce dispersed, and supplier relationships disrupted, requiring expensive restart if the design is revived. The cost of inconsistent acquisition strategy extends beyond individual programs to erode industrial base capabilities that take years to rebuild.

The Coast Guard now faces planning for NSC(X) as the eventual Legend-class replacement, potentially incorporating ice-hardened hulls for Arctic operations, integrated power systems, and enhanced capabilities for 21st-century threats.

Whether institutional memory of the NSC experience proves sufficient to avoid repeating its mistakes, inconsistent funding, strategic-budgetary misalignment, zero-sum platform competitions, remains to be seen.

Admiral Currier's 2011 testimony provides the roadmap: stable funding enables efficient production, clear strategic vision must match resource commitment, and acquisition discipline requires sustained leadership attention. The NSC program demonstrated both the promise of following this roadmap and the cost of abandoning it.

# 3

# THE FRIGATE THAT FINALLY WAS: HOW THE NAVY CAME HOME TO THE NATIONAL SECURITY CUTTER

THE U.S. NAVY spent two decades and tens of billions of dollars proving what should have been obvious from the start: a platform that was already in production, already proven at sea, and already meeting the demands of long-range maritime patrol was sitting right there in the Coast Guard inventory. The Legend-class National Security Cutter was never a secret. It was simply ignored.

By late 2025 and into 2026, the Navy had effectively ended the Constellation-class program and pivoted to an NSC-derived design as the baseline for its new FF(X) frigate. Call it a correction. Call it a belated act of common sense. What it unmistakably is, is a validation of a path that was available in 2003 and rejected in favor of transformational ambition that never transformed into operational capability.

The Australian industrialist Essington Lewis, who spent years warning his country of the coming conflict with Japan before World War II, framed it simply: "Time is more important than money for defence preparedness." The Navy's frigate saga is a case study in what happens when that lesson is ignored. Money lost can be appropriated. Time lost cannot be recovered.

## The Path Not Taken

Between 2002 and 2005, as the Navy was locking itself into the Littoral Combat Ship concept, high speed, modular swappability, ambitious unmanned integration, while the Coast Guard was taking a straightforwardly different approach. The National Security Cutter program was built around proven hull forms, mature propulsion, and mission requirements grounded in operational reality.

An NSC-based frigate was not a novel idea. The hull, propulsion plant, and aviation infrastructure were already there. The Coast Guard's built-in requirements for endurance, seakeeping, and aviation support aligned naturally with what frigates actually do, patrol, escort, presence, and combat operations at the low-to-medium intensity end of the spectrum.

Instead, the Navy chose the LCS concept because it rejected evolutionary thinking. Two competing hull forms would drive innovation. Swappable mission modules would make a single hull perform ASW, mine countermeasures, or surface warfare as required. The vision was intoxicating. The execution was close to catastrophic.

## Twenty Years in the Wilderness

The LCS failure is well documented, but the pattern bears repeating because it is the necessary context for understanding why the Navy has ended up where it has. Both LCS variants proved mechanically brittle. Galvanic corrosion, gearbox failures, and propulsion shortfalls forced speed restrictions and costly redesigns. More fundamentally, the mission modules never became what the program promised.

The mine warfare module could not reliably detect and neutralize mines at the required speed and accuracy. The ASW suite never delivered the reach or lethality to prosecute modern submarines in contested water. The surface warfare fit was under-armed against any plausible peer threat. And the modularity concept itself, the conceptual heart of the program, collapsed in practice: modules were not swapped at sea, doctrine for multi-mission employment never fully developed, and sustaining three distinct mission sets across a small hull class proved logistically unworkable.

What this meant operationally was stark: LCS could not do what frigates must do. Without credible organic air defense, meaningful anti-ship capability, or real ASW depth, LCS could not escort high-value units, screen amphibious task forces, or hold its own in a contested maritime environment. As major power competition reasserted itself, LCS became a ship without a mission that matched the threat.

The Navy began decommissioning LCS hulls well before their planned service lives while simultaneously retaining others for niche roles it could not otherwise fill. The ambivalence captured the program's essential problem: the ships did not fit the need, but the fleet could not simply walk away from the hulls it had.

The Constellation-class was supposed to fix this. A FREMM-derived design, announced in 2017, promised a traditional multi-mission frigate built around proven systems. But Americanization, requirements growth, and design instability drove up costs and stretched schedules. By the mid-2020s, lead-ship delivery had slipped years past the original 2026 target, and design stability remained elusive. Constellation had become its own version of the LCS problem, a requirement sink and a schedule drag at exactly the moment the surface fleet needed credible hulls in the water.

The decision to terminate or sharply limit Constellation and pivot to an NSC-derived FF(X) was not an act of strategic clarity. It was a belated acknowledgment that time had run out.

## Coming Home to the NSC

Against that backdrop, the choice of the NSC as the FF(X) baseline is both pragmatic and revealing. The Legend-class offers what the Navy now understands it needs: a proven hull, a functioning industrial base, a known cost profile, and the ability to move from decision to delivery on a timeline that actually matters.

At roughly 418 feet and 4,500 tons displacement, the NSC carries ample volume and weight margins to accept military systems. Its propulsion and hotel load were designed with growth in mind from the start. Its long range, sustained endurance on station, and robust aviation facilities make it a natural for escort, presence, and partner-engagement operations across the Pacific and beyond.

The industrial logic is straightforward. Ingalls has delivered NSCs to the Coast Guard on a roughly annual cadence. The workforce knows the design. The supply chain exists. Building FF(X) on a line with a real learning curve, rather than standing up a new production effort from a paper concept, reflects a hard-won understanding that schedule and throughput matter as much as technical specification in a constrained industrial base where nuclear submarines, large surface combatants, and Coast Guard recapitalization all compete for the same yard capacity and skilled labor.

## The Strategic Price of Lost Time

The temptation to celebrate the Navy's arrival at a sensible solution should be resisted until the cost of the journey is accounted for.

Had the Navy pursued an NSC-based frigate starting in 2003–2005, the fleet could today field twenty or more mature, multimission frigates with years of operational experience, established tactics, and integrated roles in carrier and expeditionary strike groups.

These would not have been theoretical ships. They would have been on station in the Gulf of Aden during the height of Somali piracy. They would have been present in the Western Pacific as China expanded its maritime footprint. They would have been available for escort, ballistic missile defense patrols, and distributed maritime operations as those concepts developed. Their existence would have shaped adversary planning and allied perceptions.

China did not squander its time. The People's Liberation Army Navy moved from a largely coastal force in the early 2000s to a substantial blue-water fleet by the mid-2020s, with hull counts surpassing the U.S. Navy. Its Type 054A frigate, more than forty hulls commissioned, entered service in 2008 and has been produced on a steady, evolutionary basis since. While the United States chased ambitious but fragile designs, China invested in repeatable ship-building and accumulated the operational experience that goes with it.

Essington Lewis understood something that the Navy's program managers did not: building industrial capacity, training the workforce, and developing operational doctrine all require time. Every year consumed by a failed or unstable program is a year not used to build hulls, train crews, and learn how to fight with the platforms you actually have. Delays compound. In a long competition, they become structural disadvantages.

The frigate gap that opened over two decades forced operational compromises that consumed the fleet's most capable assets. High-end destroyers and cruisers were diverted to escort and presence missions that frigates should have handled. Distributed operations, partner engagement, and persistent forward presence all suffered from a shortage of appropriately-sized combatants. The readiness cost was real, even if it was diffuse and difficult to attach a dollar figure to.

Even now, as the Navy pivots to FF(X), it is stitching together a patchwork: retained LCS hulls for specialized roles, Burke-class destroyers stretched thin across a global demand signal, and an NSC-derived frigate that will take years to reach meaningful numbers. The FF(X) decision is necessary and, in important respects, welcome. But it cannot unwind twenty years of underperformance.

## The Coast Guard Dimension

For the Coast Guard, the Navy's decision to come home to the NSC has concrete implications that deserve direct attention.

The white-hull service was the one that took the disciplined path in the 2000s, building the National Security Cutter and following it with the Offshore Patrol Cutter as part of a generational recapitalization. Those programs have faced their own cost pressures and schedule challenges, but the underlying logic, proven designs, evolutionary upgrades, realistic mission requirements, stands in sharp contrast to the Navy's oscillation between LCS and Constellation.

Now the same industrial capacity that built NSCs for Coast Guard missions is being asked to produce a frigate for the Navy, at exactly the moment the Coast Guard is struggling to replace aging medium-endurance cutters and bring OPCs online at scale. The demand signal for hulls, skilled labor, and yard capacity is converging across two services.

The convergence cuts both ways. On one side, FF(X) production can stabilize shipyard workloads, sustain the skilled workforce, and justify facility investments that benefit both services. A steady run of NSC-derived hulls, some in white paint, some in haze gray, could give the United States a more resilient shipbuilding industrial base than two separate small-batch programs ever could.

On the other side, competition for yard slots, materials, and engineering attention creates friction. If Navy urgency in fielding FF(X) translates into priority access to the NSC production line, Coast Guard timelines for cutters already operating beyond their intended service lives could slip further. That is not an abstract concern. It is an operational risk with readiness consequences that are already materializing.

There is also a broader strategic narrative at stake. The Navy's visible embrace of an NSC-derived platform is an implicit acknowledgment that the Coast Guard made the right acquisition choices first. The "frigate that finally was" is, at its core, a validation of Coast Guard design discipline and acquisition realism. The question for both services is whether they can now synchronize their recapitalization strategies rather than compete for the same limited industrial base.

## The Path Forward

The NSC-based FF(X) offers the Navy a genuine opportunity to rebuild frigate capacity on a foundation that actually exists. To succeed, the program has to avoid the pathologies that doomed its predecessors.

Requirements discipline is the first imperative. FF(X) should be designed to be an excellent frigate, not a mini-destroyer and not a platform for every technology the acquisition community finds interesting. The initial configuration should deliver credible self-defense, meaningful ASW and surface warfare contributions, and the ability to escort high-value units. Growth margins exist in the hull. They should be preserved for future flights, not drawn down to satisfy expanding wish lists before the first ship is even laid down.

Schedule discipline is equally non-negotiable. The fleet needs frigates in the water, not perfect frigates in briefing slides. The Navy's decision to accept modest initial armament and rely in part on containerized capabilities reflects a recognition that good enough now is strategically superior to exquisite later—particularly when "later" has a twenty-year track record of not arriving.

Production stability matters as much as design discipline. One of the quiet strengths of the Coast Guard's NSC program was its roughly predictable production tempo. Not spectacular, but consistent enough to sustain skills, suppliers, and processes. The Navy should emulate that model: set a realistic build rate, protect it from annual budget turbulence, and sustain it long enough to produce a real change in fleet capacity.

Finally, the Navy and Coast Guard need to treat the NSC lineage as a shared strategic asset rather than a competitive resource. The hull underpinning FF(X) is also the backbone of high-end Coast Guard operations. Decisions about yard investments, production sequences, and design evolution will shape both the gray-hull and white-hull futures. In a constrained industrial environment, the two services can either compete destructively for the same limited capacity or coordinate to grow capacity that both depend on.

Essington Lewis's maxim is as unforgiving today as when he coined it. Time is more important than money in defense preparedness. The Navy spent lavishly on LCS and Constellation while spending its most irreplaceable resource, time, on designs that did not deliver.

The frigate that finally was is a Coast Guard cutter in Navy paint. That is not an insult. It is a description of what sound acquisition looks like: build on what works, evolve it deliberately, produce it steadily, and get it into the hands of operators before the strategic window closes.

The window did not stay open. The challenge now is to make up for lost time, not just in steel and sensors, but in the industrial habits, operational experience, and institutional credibility that only come from actually building and operating the fleet you need.

# PART 5

---

# MARITIME PATROL AIRCRAFT AND C4ISR MODERNIZATION

Part five of the book examines how the Coast Guard's modernization of maritime patrol aircraft and C4ISR transformed an aging, over-tasked aviation enterprise into a more capable, networked force that could manage twenty-first-century "chaos" at sea rather than simply respond episodically to crises.

It traces this shift from the early 2000s Deepwater vision through the first decade of post-9/11 operations, showing how new sensors, communications suites, and mission-system integration on platforms such as the HC-130J fundamentally changed what Coast Guard aviation could see, decide, and do over vast maritime spaces.

The section underscores that this evolution occurred despite persistent budget headwinds, political micromanagement, and the constant pressure to stretch legacy platforms beyond their intended service lives.

The first chapter in Part V lays out the conceptual case for maritime patrol aircraft as "strategic enablers" rather than just search-and-rescue workhorses, arguing that C4ISR integration turned these aircraft into flying command-and-control nodes linking cutters, shore commands, and joint or coalition partners.

It explains how improved domain awareness, data fusion, and secure connectivity allowed aircrews to orchestrate complex, multi-asset operations across missions ranging from counter-narcotics and fisheries enforcement to homeland security and Arctic presence.

At the same time, the chapter exposes how DHS budget decisions and competing departmental priorities repeatedly slowed or truncated the recapitalization of the maritime patrol fleet, leaving operators to bridge the gap between strategic rhetoric and operational reality.

The centerpiece of Part V is the detailed reconstruction of the Hatteras rescue, a 2010 operation in the North Atlantic that vividly demonstrates what modernized C4ISR and airframes made possible. Operating in extreme weather, with 40-foot seas, severe turbulence, icing, and a key radar system inoperative, the HC-130J crew used a layered mix of DF equipment, weather radar, SATCOM, HF, NVGs, and FLIR to locate and protect a single sailor whose vessel was breaking up under him. The narrative walks through how the crew triangulated the distress signal without primary radar, maintained continuous visual and infrared contact in darkness, managed fuel constraints, and executed a seamless handoff to a relief aircraft and a Navy helicopter that completed the hoist. The rescue becomes a case study in redundancy, crew training, and the operational payoff of investing in mission systems rather than merely replacing airframes.

The final chapter in Part V steps back to place C4ISR modernization in a broader institutional and political frame. It contrasts the Coast Guard's hard-won gains in aviation and networking with the continuing pattern of underfunding, over-management, and episodic political attention that leaves the service perpetually improvising to meet expanding mission demands.

Drawing on interviews and earlier Deepwater-era research, the book argues that the Hatteras rescue exemplifies both the remarkable professionalism of Coast Guard aircrews and the fragility of a system that relies on their ingenuity to overcome structural neglect.

Part V thus closes by positioning maritime patrol aviation and C4ISR as central to the Coast Guard's role as a "white fleet" for chaos management, while warning that these capabilities cannot be sustained or extended without breaking the long-standing cycle of being always ready yet persistently under-resourced

# 1

# THE KEY ROLE FOR MARITIME PATROL AIRCRAFT AND C4ISR MODERNIZATION: A 2011 PERSPECTIVE

THE OBAMA ADMINISTRATION came to power at a critical juncture in global affairs. The President emphasized the importance of allied collaboration and shaping new forms of global engagement for the United States. He highlighted the need to enhance international cooperation in combating terrorism with global reach while expanding American involvement in the worldwide trading and economic system. The articulated vision was clear: building a new role for the United States in an evolving multipolar world.

No service embodies this vision more completely than the United States Coast Guard. The USCG operates at the nexus of commercial, law enforcement, and military activities. It carries a multi-mission mandate supporting homeland protection, global trade security, maritime counter-terrorism, sea-borne drug interdiction, shipping safety, and international collaboration in securing the global commons.

Yet paradoxically, even as the USCG's importance grows, the service faces the challenge of executing these critical missions with increasingly constrained resources in an era of global financial downturn.

The Administration's decisions regarding USCG capabilities present a troubling contradiction. Two cutters are scheduled for retirement from the Pacific fleet. The Office of Management and Budget seeks to cap new cutter acquisition at four vessels rather than the eight or nine the USCG requires. Maritime Patrol Aircraft numbers face significant cuts. None of these reductions align with stated goals of enhancing maritime safety, security, and custodianship unless the strategic objective is actually to diminish these capabilities.

The economic consequences will be substantial. The reduced Pacific fleet cannot provide adequate coverage for fisheries protection alone, guaranteeing billions of dollars in lost fishing commerce.

The gap between administration rhetoric and action regarding the USCG and its missions creates serious implications for security, defense, and economic interests. Effective territorial coverage, law enforcement, and security provision require adequate assets. The USCG performs admirably, but disappearing assets translate directly into diminished safety, security, and economic support for the nation.

The Department of Homeland Security's budget documents acknowledge this reality:

> *The Coast Guard provides agile, adaptable, and ready operational capabilities to serve the Nation's maritime interests. Throughout the U.S. maritime domain, the Coast Guard provides a recognized maritime presence in carrying out its safety, security, and stewardship roles. It is also the only DHS organization and Armed Service that can operate assets for both law enforcement and military purposes within and beyond U.S. territorial limits."*
>
> *This presence, supported by a military command, control, and communications network, gives the Coast Guard both prevention and response capabilities for all threats. The Coast Guard can augment forces from the local level to a national or international level of involvement, regardless of the contingency. Moreover, the Coast Guard can flow its unique capabil-*

*ities and authorities to DoD for national security contingencies. As both a military service and law-enforcement agency, the Coast Guard 'straddles the seam' separating the federal government's homeland-security and homeland-defense missions.**

These are excellent statements of principle, but where are the resources to execute them?

## The Connectivity Challenge

Providing this flexibility with limited resources in a constrained budgetary environment demands highly connected forces built around capable multi-mission assets. Connectivity enables effective knowledge-based action, while multi-mission assets execute based on that knowledge across varied circumstances.

Without physical assets operating in an enhanced connected environment, improved connectivity cannot translate into effective action. The ability to cover territory with aviation assets working alongside surface vessels and shore-based capabilities remains central to success for the maritime security and safety team, with the USCG serving as the crucial anchor.

The maritime patrol aircraft, including C-130s and helicopters, form a surveillance, lift, and action team that significantly extends the reach of the USCG and joint surface assets. The new platforms being introduced into the USCG, upgraded C-130s and network-enabled helicopter assets, substantially enhance the capability to work with joint teams. In an era of financial stringency, making effective use of joint assets across the USN, US Air Force, and USCG becomes essential. The new connectivity-enabled air assets make this possible.

------

*  https://tile.loc.gov/storage-services/master/gdc/gdcebookspublic/20/23/69/53/11/2023695311/2023695311.pdf

However, this integration cannot occur without the platforms themselves and their modern integrated systems. Enhanced connectivity is not a substitute for platform acquisition; rather, it serves as a force multiplier for new network-enabled platforms.

## The Connectivity Enhancer

Before the September 11 attacks, the USCG developed a new acquisition approach called Deepwater, designed to dramatically recapitalize aging physical assets with limited resources. The acquisition of surface, shore, and air assets would integrate with common command, control, communications, computers, intelligence, surveillance, and reconnaissance (C4ISR) tools to provide interoperability across the range of maritime safety and security stakeholders.

The goal centered on C4ISR systems, largely rooted in commercial and government off-the-shelf procurement, as the backbone for an integrated service. A new central nervous system would enable fewer but more capable physical assets. The events of September 11 accelerated this approach and demanded new capabilities. Deepwater was redirected by emerging requirements as the USCG faced newly defined maritime security threats. The evolutionary approach was disrupted by surge requirements to address immediate threats.

Deepwater critics focused on problems with certain asset acquisitions, particularly patrol boats and, to some extent, the new National Security Cutter. The USCG responded by creating a new acquisition directorate to subsume Deepwater.

Lost in this shuffle was the significant success of several shore and aviation assets, the core C4ISR systems, and efforts to provide interoperability with civil, commercial, and military partners. The addition of a classified network, upgraded Inmarsat communications, deployment of an automatic identification system, upgraded law enforcement radios, and other systems greatly improved USCG capabilities.

The impact of these new capabilities appears clearly in various

USCG missions, including drug trafficking interdiction. The Coast Guard combats drug trafficking in the Caribbean, Gulf of Mexico, and Eastern Pacific, accounting for more than 50 percent of all cocaine seizures annually. The service also stops thousands of illegal migrants, avoiding billions of dollars in social services costs while aiding persons and vessels in distress, saving lives and minimizing injuries.

Accomplishing these challenging missions requires technology: electronic charts integrating with radar, computerized planning and tracking systems, infrared-detection gear, all-weather day-night cameras, and secure channel radios.

Multi-mission stations, cutters, aircraft, and boats linked by communications networks permit effective operations. Electronic systems enable coordinated tactics, integrated intelligence, multi-agency interoperability, and common situational awareness necessary for fulfilling statutory missions.

The continued increase in cocaine seizures provides a good measure of technology implementation success. New communications systems, coupled with the latest electronic sensor technology, make drug trafficking intelligence rapidly available and improve target tracking, achieving effective prosecution of criminal activities.

These systems, deployed across multiple asset types, allow for equipment interoperability and software commonality. Operators can transmit and receive classified and unclassified information to and from other assets, including surface vessels, aircraft, local law enforcement, and shore facilities. The electronic network serves as the glue permitting collaborative achievement of common purposes, a true force multiplier.

A common architecture deployed across multiple asset types allows for equipment and software system commonality and supportability across the entire Coast Guard enterprise. Generally, the Deepwater C4ISR architecture ensures an open systems approach for design and implementation, providing a web-enabled infrastructure. Its architecture adapts to technology insertion and

enables progression to future Coast Guard-wide C4ISR architectures.

The first National Security Cutter, USCGC Bertholf, passed all
TEMPEST and information assurance requirements, culminating in
authority to operate. Lockheed Martin, one of the Deepwater partners, received specific recognition in a Coast Guard letter of appreciation dated June 24. Delivery of the second National Security
Cutter, USCGC Waesche, also received an excellent C4ISR evaluation, with the ship accepted with no C4ISR trial cards. Systems
include electronic charts integrating with radar, computerized planning and tracking systems, infrared-detection gear, all-weather day-
night cameras, and secure channel radios.

## Multi-Mission Platforms
## and the C4ISR Advantage

The Coast Guard has accepted delivery of missionized HC-130J
Long Range Surveillance Maritime Patrol Aircraft. The aircraft's
new mission equipment and sensor packages deliver enhanced
search, detection, and tracking capabilities for maritime search and
rescue, maritime law enforcement, and homeland security missions.
Aircraft modifications include installation of belly-mounted surface
search radar, nose-mounted electro-optical infrared sensors, flight
deck mission operator stations, and mission-integrated communication systems. The mission system installed on the HC-130J derives
from the same software series developed for the mission system pallet
aboard the HC-144A maritime patrol aircraft.

The Coast Guard has also accepted mission system pallets for
their new HC-144A "Ocean Sentry" Medium Range Surveillance
Maritime Patrol Aircraft. This roll-on, roll-off suite of electronic
equipment enables aircrews to compile data from multiple integrated
sensors and transmit and receive both classified and unclassified
information to other assets, including surface vessels, aircraft, local
law enforcement, and shore facilities.

At the heart of all these platforms lies a common command and control system providing system interoperability and commonality for maintenance and training savings. C4ISR system software reuse across Coast Guard platforms is significant. The same 100 percent common software between air platforms is also 41 percent common with software on the National Security Cutter. The suite on the National Security Cutter leverages nearly 75 percent of the code from the Navy's latest Aegis baseline, while the air suite shares 80 percent commonality with the Navy's P-3 Aviation Improvement Program. The command and control system is compatible across the Coast Guard's land, air, and sea assets while remaining interoperable with the Navy and 117 federal agencies.

In ports and coastal areas, one of the USCG's most significant capability enhancements is shaping a robust connectivity capability for the USCG and its maritime and security partners. It represents a fundamental building block in improving the Coast Guard's ability to maintain maritime domain awareness focused on meeting the needs of decision makers engaged in operations at sea, ashore, and in the air.

Despite such developments, the Coast Guard has insufficient new assets. Much of the fleet wears out at an ever-increasing rate. Modern communications and electronic systems are needed so that new and aging assets can deploy effectively. The aviation assets are not being modernized at a sufficient rate to enable the USCG to have the reach necessary to act upon newly acquired information generated by new C4ISR systems.

## The Maritime Patrol Aircraft
## at the Center of the Enterprise

When considering a maritime service, one typically thinks primarily of ships. Yet in a 21st-century maritime enterprise, data, communications, and integrated air and surface systems truly lie at the heart of effective operations. The Administration seeks within the Department of Defense to organize more effective integration of the USAF with the USN. At the heart of this new air-sea approach lies core recognition of the role of air and surface asset integration in mission success.

The USN has underscored the central role its new maritime patrol aircraft and unmanned vehicles (MMA and BAMS) will play in maritime operations. Illustrative of this connectivity contribution by the USN's aviation fleet, Rear Admiral Brooks noted:

> *The MMA will be a rapidly deployable force capable of operating independently in the far reaches of the world's oceans or in coordination with naval, joint, or allied forces. Its superior connectivity ensures MMA a major role in FORCEnet as a key airborne node. MMA's projected state-of-the-art communications and data links have the potential to act as sensors and provide the airborne node necessary for the Navy's network-centric C4ISR architecture of the common undersea and common operations pictures. This capability will enable real-time, multi-channel command, control, communications, and intelligence links combined with onboard sensing, quick-look analysis, data fusing, and data dissemination to deliver timely information to naval, joint, or allied decision makers.**

U.S. forces discovered in Iraq and Afghanistan the central role aviation assets, manned and unmanned, play for successful ground force operations. The USMC concept of the MAGTF, the combined air and ground force, is being replicated to a considerable extent as the U.S. Army works with the U.S. Air Force in shaping joint concepts of operations.

The USCG uses its aviation assets similarly. Aviation assets, helicopters and MPAs, exist as extensions of ground and surface fleet capabilities. For the 21st-century USCG, much like for the 21st-

---

* https://sldinfo.com/2011/04/shaping-a-21st-century-uscg-the-key-role-for-maritime-patrol-aircraft/

century U.S .military overall, air assets extend the reach, range, and capability to act of surface assets. For the USCG, this means a simple truth: a surface ship without the long reach of an MPA or a UAV cannot see very far. By extending the sight and reach of a ship or fleet, the ability to act and protect US equities and interests expands significantly.

With the addition of multi-mission pallets to the MPAs and the ability to integrate information gained by those systems within shipboard operations and decision-making, notably with the new USCG cutter, the MPAs become especially significant extenders of fleet activity.

Without this extended reach, drugs may well enter the U.S., illegal immigrants may go undetected, and illegal shipping with whatever consequence may enter U.S .waterways. Inadequate resources lead to inability to acquire assets with clear consequences for reduced USCG ability to see and act on knowledge to protect the US homeland or forces operating abroad.

The new MPA's concept of operations is essential to the functioning of the 21st-century USCG. The new MPAs are significant enablers for the legacy and modernized fleet. The best way to visualize this is a simple Venn diagram in which the surface asset occupies one circle, and the MPAs provide a second circle significantly expanding the range of operation and hence the ability to execute missions.

The concept of operations typically depends on the mission and is essentially the same as the HC-130. Part of this calculation involves determining proper track spacing (distance between search legs), which is a function of target size (a man in the water would require very small track spacing), sea state, visibility, and cloud cover. Night searches are conducted if the target is thought capable of detection by radar or possesses radio or signaling equipment. Typically, a ship controls one or more MPA/helicopters, continuously updating which areas have been searched and the results. This function can also be performed by a shore station. The primary reason for ship involvement is to be available as a helicopter platform and to launch a boarding party, engage the target if hostile, or recover survivors.

However, the HC-144 can do more than just pass target position information or provide vectors. It has the ability to take photos and download them, so the ship can have information on the number of people aboard the target craft (valuable intelligence if they are hostile) and perhaps even information on arms.

Finally, fixed-wings are often required to loiter over targets or follow them until a surface asset can intercept. Endurance is critical here and represents a major shortcoming of the HU-25. The HU-25 has about four hours of endurance regardless of altitude. It can fly much farther at high altitude than low altitude, but four hours is the limit. The HC-144 can provide ten hours.

## Operational Examples

The role of the MPA as a force enhancer appears clearly in several examples. In late July 2009, the crew of the National Security Cutter Bertholf conducted the cutter's first drug bust and disrupted a major drug smuggling operation in international waters. Two suspected drug smuggling boats, four suspected smugglers, and a bale of cocaine were seized as evidence some 80 miles off the coast of Guatemala.

The incident began when a group of four suspicious "pangas" were spotted by a maritime patrol aircraft, which alerted the Bertholf. A Coast Guard helicopter launched from the cutter, and a marksman aboard was able to shoot out the engines of two speedboats and fire warning shots at the other two during the pursuit as bales were being thrown overboard from all four boats. Shortly after, Bertholf's interceptor boats apprehended two of the boats and detained the four people onboard.

The concept of operations described in this operation is typical of an integrated air-to-surface operation. Obviously, all of the assets described are necessary for mission success, but the MPA as spotter and tracker is clearly at the heart of such success.

An additional example underscores the significance of the USCG's ability through its new connectivity solutions to work with other U.S .aviation assets to achieve mission success. In a 2008 drug seizure in the Pacific, a USN MPA was central to mission success. USCG Commandant Allen wrote:

*I'm proud to tell you that over the past five days, Pacific Area Coast Guard units, with the help of our U.S. Navy and interagency partners, seized over 14 tons of cocaine from two Self-Propelled Semi-Submersible vessels in the Eastern Pacific Ocean. On September 13, a Navy maritime patrol aircraft detected an SPSS in international waters and vectored a Navy ship, with a Pacific Area Law Enforcement Detachment, to intercept it. After conducting an unannounced nighttime boarding, the LEDET discovered 7 tons of cocaine. Then on September 17, another Navy MPA detected an SPSS and vectored the cutter Midgett to investigate. Midgett's boarding team subsequently discovered an additional 7 tons of cocaine. The interoperability between Coast Guard and Navy assets has never been more effective.* [*]

MPAs are not only central to extending the reach of surface ships but as multi-mission assets can operate independently to provide for independent actions, ranging from C4ISR missions on their own for the benefit of DHS decision-makers to lift assets able to move goods and personnel as they did in dealing with the aftermath of Hurricane Katrina.

With flood waters from the Mississippi River threatening several Midwestern states in June 2008, the U.S. Coast Guard deployed one

---

[*] https://sldinfo.com/2011/04/shaping-a-21st-century-uscg-the-key-role-for-maritime-patrol-aircraft/

of its newest aircraft to survey the damage and assess the condition of vital shipping lanes. The HC-144A Ocean Sentry flew over flooded portions of eastern Iowa and northeast Missouri on June 19, 2008. Though its initial operational capability was still six months off, Rear Admiral Joel Whitehead, Commander Coast Guard District Eight, selected the HC-144A because he believed this fixed-wing, turbo-prop aircraft's longer ranges, slow flying speeds, and passenger capacity best suited the mission needs. Also affecting its selection over other Coast Guard aircraft was the limited availability of HU-25 Falcon jets, the aircraft the HC-144A will eventually replace.

The mission took senior Coast Guard officials, including Rear Admiral Whitehead and maritime industry representatives, on a flight to observe the extent of flooding. The shipping industry representatives flew along to gain better understanding of how the floods would impact shipping and commerce along the Mississippi River and its tributaries in Iowa and Missouri. The Ocean Sentry's ability to fly at slower speeds allowed for more effective aerial assessment and increased time on scene.

In crises like the Haiti rescue effort, the lift and logistical capabilities of the USCG become central to the national response. Although UAVs flying over the island provide useful intelligence, better information means little without the lift and logistical capability of aircraft like the USCG C-130s and MPAs.

## The Challenge of Moving Forward

The legacy fleet is in extremis, and the need to provide new aircraft built around significant improvements in maintenance costs is essential to ongoing USCG operations. The Coast Guard initially bought 41 HU-25A aircraft, of which 36 were operational. The operational fleet size was reduced in the early 1990s by Admiral Kime to fund expansion of the Marine Safety Program. The very high cost of maintaining the HU-25, especially the engines, has been a significant burden on USCG operations.

Currently the HU-25 fleet is being reduced as HC-144s are brought online. The current fleet is only 18 aircraft: 14 operational and 4 support aircraft. The Coast Guard has made minimal investment in the HU-25 over the past several years to save money. For example, the aircraft is limited to FL230 (service ceiling is FL420) because the Coast Guard elected not to invest in an upgraded IFF several years ago.

The new USCG MPAs are the lead acquisition in the Deepwater program. The Deepwater program was built around USCG assessments of an integrated approach to executing core USCG missions. This assessment was shaped in the mid-1990s and then recast in the wake of post-September 11 assessments. The core approach held that an integrated C4ISR capability would allow the USCG to avoid a costly one-for-one replacement approach for its obsolescent assets. By integrating the replacement assets, fewer numbers of each particular asset would be required.

Although the execution of the Deepwater program was flawed, the underlying approach remains sound. Yet C4ISR funding for the USCG is on a downward trajectory, and the numbers of replacement assets are going down, not up. This is the worst of all possible worlds. The C4ISR, which enabled the acquisition of fewer platforms, is underfunded, and the result is not more platforms being acquired but fewer.

The only result of this will be a significant shortfall in the USCG's ability to perform its missions, with security shortfalls and various negative consequences such as more drugs entering the United States, more illegal immigrants with possible health hazards going unnoticed, more arms trafficking from sea successfully entering the United States, and so on.

Failure to recapitalize the USCG is not merely a funding game; it is a pulling apart of the network that protects the United States. Holes in the net are pathways for forces creating negative consequences on U.S. territory.

In the original Deepwater schema, the USCG was projected to

acquire 35 new MPAs and 6 high-altitude endurance organic UAVs. The UAVs were cancelled, although the USCG and DHS are exploring ways to supplement or replace the cancellation. With this cancellation, organic airborne ISR was reduced, and logically, the number of MPAs should increase by some function of the 6 cancelled UAVs, suggesting the number might now approach 40 MPAs.

The program is on track and on budget. The HC-144A Ocean Sentry is a derivative of the European Aeronautics and Defense Systems (EADS) CASA CN-235-300 twin-turbo medium cargo/transport aircraft. It is fitted to carry sensors, search radar, and an advanced communications suite. Forming the Ocean Sentry's nerve center is the Mission Systems Pallet (MSP) that integrates all the aircraft's systems with a two-place operators' station inside the cargo compartment. The MSP is a roll-on/roll-off suite of electronic equipment that allows aircrews to compile data from the aircraft's multiple sensors and transmit that data to other participants in the operational area.

As the USCG describes the new MPA:

*This fixed-wing turbo prop aircraft provides invaluable on-scene loitering capabilities and performs various missions, including maritime patrol, law enforcement, Search and Rescue, disaster response, and cargo and personnel transport. The Mission System Pallet is a roll-on, roll-off suite of electronic equipment that enables the aircrew to compile data from the aircraft's multiple integrated sensors and transmit and receive both classified 'Secret'-level and unclassified information to other assets, including surface vessels, other aircraft, local law enforcement and shore facilities. With multiple voice and data communications capabilities, including UHF/VHF, HF, and Commercial Satellite Communications, the HC-144A will be able to contribute to a Common Tactical Picture and Common Operating Picture through a networked Command and Control system that provides for data sharing* *via SATCOM. The aircraft is also equipped with a vessel Automatic Identification System, direction finding equipment, a surface search radar, an Electro-Optical/Infra-Red system, and Electronic Surveillance Measures equipment to improve situational awareness and responsiveness.*[*]

The new MPA is the product of a blend of several mature technologies produced by the best defense and aerospace companies in the world. In other words, it is a mature product in initial production with no problems anticipated in either continued production or ramp-up of production.

Starting with delivery in 2003, eight new MPAs were delivered through the end of 2008 along with three mission pallets. MPAs nine through 11 were scheduled for delivery in 2010 along with mission system pallets four through twelve. The MPA baseline fielding plan called for 16 additional MPAs to be delivered by 2014 for a total of 36.

Unfortunately, senior DHS acquisition officials have reportedly delayed the program. The proposal is to have a break in production after the deliveries in 2010. Such a delay to save near-term acquisition monies will come at the expense of needed capabilities. An interruption in production will not be matched by an interruption in the threat environment facing the United States or in the normal challenges facing the USCG. A funding delay is a further opening in the security net around the United States. An interruption in production is penny-wise and pound-foolish.

To quote the DHS budget document:

*The Coast Guard embraces a culture of response and action, with all of its personnel trained to react to 'All Threats, All Hazards.' In many cases, front-line operators are encouraged to*

---

[*]   https://sldinfo.com/2011/04/shaping-a-21st-century-uscg-the-key-role-for-maritime-patrol-aircraft/

*take action commensurate with the risk scenario presented, without needing to wait for detailed direction from senior leadership. This model enables swift and effective response to a wide variety of situations.*[*]

But it is difficult to provide "swift and effective response" with missing capabilities. The absence of capabilities means simply put an enhanced probability of threats not being met today and augmented tomorrow.

## Conclusion

The maritime patrol aircraft represents far more than just another platform in the USCG inventory. It serves as a critical enabler, force multiplier, and network node that transforms the effectiveness of the entire maritime security enterprise. The integration of modern C4ISR systems with capable airframes creates a capability greater than the sum of its parts, extending the reach and effectiveness of surface assets while providing independent mission capability when required.

The strategic vision articulated by the Obama Administration for enhanced global engagement, allied collaboration, and homeland security requires the very capabilities that budget reductions threaten to eliminate. The gap between strategic rhetoric and resource allocation creates real vulnerabilities in maritime domain awareness, drug interdiction, illegal immigration control, and general maritime security.

The success of the Deepwater C4ISR integration, demonstrated through numerous operational successes, proves the validity of the network-centric approach. However, this success depends absolutely on having adequate numbers of modern platforms to leverage the

---

[*]  https://tile.loc.gov/storage-services/master/gdc/gdcebookspublic/20/23/69/53/11/2023695311/2023695311.pdf

connectivity advantages. Reducing platform numbers while maintaining or reducing C4ISR funding represents the worst possible combination, neither the platform-centric approach of the past nor the network-enabled approach of the future, but instead a capability gap that leaves the nation vulnerable.

The HC-144A Ocean Sentry and upgraded HC-130J represent mature, proven technologies delivered on schedule and on budget. They embody the successful integration of commercial and military technologies, international partnerships, and modern acquisition practices. Interrupting their production to achieve short-term budget savings sacrifices long-term capability and increases overall program costs.

The United States faces evolving maritime threats that demand persistent surveillance, rapid response, and the ability to operate across vast ocean areas. Maritime patrol aircraft provide these capabilities in ways that surface vessels alone cannot match. The question is not whether the nation can afford to acquire and maintain an adequate MPA fleet, but whether it can afford not to. The costs of inadequate maritime security, in terms of drugs entering the country, illegal immigration, lost fisheries revenue, and compromised homeland security, far exceed the investment required to maintain capability.

The path forward requires alignment of resources with stated strategic priorities, recognition of the central role of aviation in maritime operations, and commitment to completing the recapitalization of aging assets before they fail. The USCG's unique position at the intersection of law enforcement, military operations, and homeland security makes it indispensable to national security. Maritime patrol aircraft are indispensable to the USCG. Therefore, adequate MPA acquisition and modernization must be recognized as a national security priority deserving of sustained funding and support.

April 19, 2011

# 2

# THE USCG GETS IT RIGHT: THE 13TH OCEAN SENTRY AIRCRAFT ARRIVES AHEAD OF SCHEDULE

IN A 2012 ARTICLE, I highlighted the delivery of the 13th Ocean Sentry Aircraft. The planes were produced by Airbus in Seville, Spain from a mature assembly line. And the process worked well.

I visited the final assembly line in Seville in 2010 and 2013 and met with personnel and reviewed the progress. This was a case of working with a competent company delivering a core product that they produced for customers worldwide.

Notably, several of the Latin American countries with whom the USCG worked also flew the same CASA/Airbus aircraft.

According to the Airbus Military press release released at the time:

*Airbus Military, via the prime contractor EADS North America, has delivered the 13th HC-144A Ocean Sentry maritime patrol aircraft to the U.S. Coast Guard from its final assembly line in Seville, Spain, two months ahead of the contractual delivery date. The HC-144A is based on the Airbus Military CN235 tactical airlifter, more than 250 of which are operated by 27 countries.*

*The Ocean Sentry is replacing the Coast Guard's aging fleet of HU-25 "Guardian" jets and some older HC-130H aircraft, and has seen service in a wide variety of missions since achieving initial operational capability with the Coast Guard in 2008. Earlier this month, an HC-144 Ocean Sentry crew located a private aircraft that had gone down in waters off the coast of Andros, Bahamas, and provided support for a successful rescue of the two passengers by Coast Guard helicopter.*

*In addition to search and rescue, the Coast Guard is utilizing the Ocean Sentry's superior  endurance and flexibility for missions including maritime patrol, cargo and personnel transport, intelligence/surveillance/reconnaissance, and disaster relief.*

January 30, 2012

Here are a couple of photos I took when visiting the final assembly line in 2010 which highlighted their building of the Ocean Sentry aircraft for the USCG.

# 3

# THE IMPACT OF MODERNIZATION: A HATTERAS RESCUE OPERATION

THE COAST GUARD that my work with Admiral Gilbert revealed through countless hours of interviews with coasties from every rank and specialty was a service defined by an almost stubborn dedication to excellence, even as it operated under conditions that would have crippled a less committed organization.

The politicians spoke of readiness. The admirals testified about modernization. But the men and women we interviewed told a different story: one of aging equipment held together by ingenuity, of missions accomplished despite rather than because of the resources provided, of a service culture so deeply committed to saving lives and protecting the nation that it refused to let political neglect become an excuse for operational failure.

The Hatteras rescue operation of January 2010, detailed in the accompanying interview, crystallizes everything I learned during those years with Admiral Gilbert.

It is a story that deserves to be understood not merely as an impressive feat of seamanship and airmanship, but as a window into the reality of a service that has been systematically underfunded and over-managed by politicians who neither understand nor apparently care about the consequences of their decisions.

On the night of January 3, 2010, Lieutenant Commander Lacer Driver and his crew launched from Elizabeth City in their HC-130J Hercules to respond to an Emergency Position Indicating Radio Beacon signal from a distressed vessel 280 miles east of Cape Hatteras.

What followed over the next several hours was a rescue operation conducted in conditions that pushed both crew and aircraft to their absolute limits, sustained 60-knot winds, 30-to-40-foot seas, solid overcast from 800 feet to 13,000 feet, icing, severe turbulence, snow squalls, and darkness.

The sailor they were seeking had been at sea for five days. His vessel had been knocked down repeatedly. His mast was gone, his engine dead, his radio failing. He was alone in the North Atlantic in winter, in a boat that was literally coming apart beneath him, being beam-to massive seas he could no longer control.

The Coast Guard found him. They stayed with him for four hours in conditions that would ground most aircraft. They delivered rescue rafts to him just moments before fuel exhaustion forced them to depart. When a second massive wave rolled his vessel and threw him into the sea as his boat sank, he surfaced, swam a few strokes, and collided with one of those rafts in the darkness. Minutes later, a Navy helicopter from the USS Eisenhower hoisted him to safety.

It was, by any measure, miraculous. But it was not a miracle. It was the product of exceptional people using advanced systems under the most challenging conditions imaginable.

And here is what Washington refuses to understand: the new HC-130J aircraft and its integrated mission systems made this rescue possible. The old HC-130H that the Coast Guard had been flying for decades could not have accomplished this mission.

The technological gap between the old and new aircraft was not marginal. It was the difference between success and failure, between life and death.

During my years of interviews with Admiral Gilbert and countless Coast Guard personnel, one theme emerged with painful clarity: the service had been operating legacy equipment far beyond its designed service life, and the consequences were measured not just in maintenance costs and operational limitations, but in missions that could not be accomplished and lives that could not be saved.

The Hatteras rescue illustrates this with brutal precision. At 70 miles offshore, the HC-130J's DF430 direction-finding system locked onto the EPIRB signal. LCDR Driver was emphatic in the interview: with the old HC-130H systems, they would have needed to be "much, much closer" to acquire any signal at all. In those conditions, in that weather, those extra miles and extra minutes were everything.

The difference between old and new technology was the difference between finding a dying man in time and arriving too late.

Consider what that crew had available to them on the HC-130J: advanced direction-finding equipment that could triangulate a beacon from unprecedented distances; integrated night vision systems that allowed them to see the vessel and the sailor's flashlight in pitch darkness; Forward Looking Infrared Radar that let them monitor the mariner's movements on deck without demanding he use his failing radio; satellite communications that enabled real-time coordination with Elizabeth City, with the relief aircraft, and with the Navy helicopter; computer systems that calculated, even if the extreme conditions rendered them ultimately unusable—sophisticated drop patterns for the rescue rafts.

And here is the crucial point that politicians consistently fail to grasp: even with one major system down (the Multi-Mode Radar was inoperative that night), the redundancy and integration of the HC-130J's mission systems allowed the crew to adapt and overcome. They used the DF430 for triangulation when they couldn't use radar. They used FLIR and night vision goggles in concert to maintain visual contact. They used multiple radio systems, VHF-FM, Inmarsat satellite phone, and HF radio, to coordinate a complex multi-platform rescue involving three aircraft across hundreds of miles.

This is not equipment for equipment's sake. This is not bureaucratic empire-building or defense contractor boondoggling. This is the difference between operational capability and operational fantasy. And the Coast Guard had to fight for every dollar to replace aircraft that were, in some cases, older than the pilots flying them.

What Admiral Gilbert helped me understand during those years was that the underfunding of the Coast Guard was not a simple budget issue. It was a systematic failure of political imagination and courage.

The Coast Guard operates across an impossibly broad mission set: search and rescue, maritime law enforcement, drug interdiction, illegal immigration control, port security, aids to navigation, icebreaking, environmental protection, maritime safety regulation, and, increasingly, defense operations in both traditional and gray-zone conflicts.

It is the only military service that operates under the Department of Homeland Security in peacetime, yet it is expected to seamlessly integrate with the Department of Defense during conflict. It protects 95,000 miles of coastline and 3.4 million square miles of ocean. It is responsible for the security of 360 ports. And it does all of this with a budget that represents a rounding error in Pentagon spending.

Politicians love the Coast Guard in the abstract. They praise coasties at medal ceremonies and in campaign speeches. They express profound gratitude for the service's heroism during hurricanes and its vigilance against drug trafficking. But when budget time comes, the Coast Guard is the service nobody fights for. It lacks the constituency of the Navy with its shipyards in key congressional districts. It lacks the aviation lobby that supports the Air Force. It lacks the political power to protect itself.

The result is a service that has been chronically starved of resources while simultaneously being burdened with expanding missions and increasing operational tempo.

During my interviews with Admiral Gilbert, we met maintainers who performed daily miracles keeping aircraft flying that should have been retired years earlier. We spoke with boat crews operating in the Bering Sea in cutters that needed extensive repairs before each patrol. We documented pilots flying more hours per year than their counterparts in other services because there simply weren't enough aircraft to distribute the mission load.

And through it all, the coasties maintained their standards. They insisted on excellence because lives depended on it. This is what made the service both inspiring and infuriating to study—inspiring because of the dedication, infuriating because that dedication was being exploited by a political system that took it for granted.

The underfunding was matched by what Admiral Gilbert called "over-management", the toxic combination of micromanagement from political appointees who had never served a day at sea and the imposition of bureaucratic processes designed for peacetime garrison operations onto a service that operated 24/7/365 in some of the most dangerous environments on Earth.

Coasties described an acquisition process so byzantine that it took years to purchase equipment that could be bought commercially in weeks. They detailed reporting requirements that consumed hours that should have been spent on training or maintenance. They explained how political sensitivities around budget justification meant that every operational decision had to be filtered through layers of Washington bureaucracy concerned more with congressional testimony than operational effectiveness.

The Hatteras rescue occurred despite this system, not because of it. The HC-130J that LCDR Driver flew that night existed because Coast Guard leadership had fought for years to replace the aging HC-130H fleet, enduring skepticism from Congress about whether the service really needed new aircraft, whether the technology improvements were worth the cost, whether the Coast Guard was exaggerating its capability gaps.

The answer, as the Hatteras rescue demonstrated, was yes,

emphatically yes. The new aircraft was not a luxury. It was not gold-plating. It was the minimum necessary equipment to accomplish missions that had become impossible with the old fleet.

And even then, the Coast Guard received fewer new aircraft than it actually needed, forcing the service to extend the service life of the old aircraft even further while waiting for sufficient HC-130J numbers to retire the legacy fleet.

What struck me most forcefully during my years of Coast Guard research was the profound professionalism that permeated the service at every level. This was not the result of generous funding or political support. It was, if anything, a cultural response to the lack of those things, a determination that if the nation would not give the Coast Guard what it needed, the coasties would make do with what they had and still accomplish the mission.

Read the interview with LCDR Driver and Petty Officer Lee carefully. Note the calm precision with which they describe operating in conditions that would justify scrubbing the mission, 60-knot winds, 40-foot seas, severe turbulence, icing, multiple system challenges. Note how they methodically worked through each problem: radar inoperative, shift to direction-finding triangulation; computer drop solutions unusable in extreme winds, revert to visual drops using night vision; vessel aspect unclear in darkness, use FLIR to determine orientation; fuel becoming critical, coordinate seamless handoff to relief aircraft while simultaneously guiding Navy helicopter to the scene.

This is decision-making at the highest level, executed under extreme stress, with life-or-death consequences, by professionals who trained for exactly these scenarios despite operating with equipment and resources that made such training more difficult than it should have been.

And here is what the politicians never grasp: this level of professional excellence cannot be maintained indefinitely in the face of chronic underfunding and over-management. Equipment that breaks too often erodes confidence. Training time consumed by bureaucratic reporting requirements degrades skills. Retention problems caused by low pay and poor quality of life conditions drain institutional knowledge. The Coast Guard's culture of excellence has been resilient, but resilience is not infinite.

The Hatteras rescue is not unique. During my work with Admiral Gilbert, I documented dozens of similar operations, rescues accomplished against incredible odds, law enforcement operations that interdicted significant drug shipments despite aging patrol boats, port security operations maintained with too few people and too little equipment, environmental responses managed with insufficient resources.

Each of these operations represented the same pattern: exceptional people accomplishing essential missions despite systematic neglect. And each represented a near-miss, not in the sense that the mission failed, but in the sense that success depended on variables that should not have been variable. Success depended on good fortune with equipment reliability. Success depended on weather being slightly better than worst-case. Success depended on problems occurring within range of available assets rather than just beyond that range. Success depended on crews being able to innovate solutions to problems that proper resourcing would have prevented.

This is no way to run a national asset as critical as the Coast Guard. The service operates in the maritime domain that carries 95 percent of U.S. international trade, through which any maritime terrorist threat would arrive, across which any mass migration crisis would travel, and in which near-peer competitors are increasingly active.

The Coast Guard is not a luxury service performing marginal missions. It is a frontline defense and security service whose importance will only grow as great power competition extends into the maritime domain.

The Hatteras rescue proves what adequate modernization enables. It demonstrates the operational value of integrated mission systems, advanced sensors, reliable communications, and well-maintained aircraft.

But it also reveals what happens when these systems are provided in numbers insufficient to the mission set and in timeframes extended by political penny-pinching and bureaucratic sclerosis.

On that January night in 2010, everything worked: the crew's skill, the aircraft's systems, the coordination with Navy assets, the luck of the survivor finding the raft in the darkness. But professional excellence should not require luck. National security should not depend on making do. Lives should not hang on whether aging equipment happens to work on the particular day when it is needed most.

Admiral Gilbert spent his career trying to make Washington understand this. The coasties I interviewed had long since stopped expecting understanding: they simply continued doing the job, maintaining the standards, saving the lives. This chapter documents one of those saves.

But it should be read as something more: as evidence of what is possible when the Coast Guard receives the tools it needs, and as an indictment of a political system that too often denies those tools while expecting miracles anyway.

The sailor pulled from the North Atlantic that night survived because the Coast Guard had managed, despite everything, to acquire new aircraft with the systems necessary to find him in time.

- How many others might have been saved over the years if the service had been properly resourced?
- How many more will be lost if the pattern of underfunding and over-management continues?

These are not rhetorical questions. They are the questions Admiral Gilbert taught me to ask. They are the questions that should haunt anyone who reads the account that follows of professionals performing at the absolute peak of their capabilities to compensate for a political system performing far below the minimum acceptable standard.

The Hatteras rescue was a triumph. But it was a triumph that should never have been necessary, accomplished by a service that should never have to fight so hard for the resources to do its job. That is the real story.

That is what Admiral Gilbert wanted the nation to understand. That is why this rescue, remarkable as it is, matters beyond the life of one sailor on one night in the North Atlantic.

Dissecting the USCG Hatteras Rescue Operation

March 9, 2010

Early January 2010, a dramatic rescue effected by the USCG was done in cooperation with the U.S. Navy to save a man's life at sea. This rescue involved key C4ISR assets on the new USCG MC-130Js.

In a time of budgetary crisis, the USCG is still in dire need for new assets as its old ones are retired because they are just too old to operate. The new ones provide significant enhancements for operations as well. Too often when a service asks for a new asset, the need seems unclear or the impact of the new over the old is also unclear.

In this article I "reverse engineered" a USCG operation to provide readers with a better understanding of how the USCG achieves success, what training, manpower and assets are required to do so.

The impact of modernization on operations could not be clearer than in this case of rescuing a sailor at sea. Not only did the new mission systems and C4ISR assets play a central role in allowing the USCG professionals to save this man's life, it would have been unlikely that either the effort or the joint team communication involved in the effort would have been possible.

Here the redundancy of the new mission systems allowed the crew to operate in extremely challenging conditions. The systems allowed them to locate the man miles before they would have been able to with the old systems; time was of the essence and the new systems allowed that time.

And when the MC130J became mission critical with regard to fuel, they were able to hand off through the C4ISR systems, the data that they had generated to the replacement MC130J. This allowed for continuity at a crucial moment.

Lt Commander Lacer Driver and his mission systems officer, AET2 S. Lee, explained in an interview on February 16, 2010 at the Elizabeth City USCG Air Station how they conducted the operation.

Without the tools available on the MC130J they would not have succeeded; without their skill, rapid decision-making process and professionalism under duress, no amount of equipment would have mattered. Crews (men and women) and the equipment are essential to this type of success.

LCDR Driver and AET2 Lee during the interview.

**Question: Why don't you explain the mission on**

## January 3rd and the challenging kind of rescue you had to do and how that operation was conducted?

Lt. CDR Driver: *Yes sir: we were the search and rescue team that night for the C130J community, stationed at Elizabeth City. We had come on duty that morning at 7:45, we flew an afternoon training flight a little over 2 hours in duration, wrapped that up and we were all just calmly waiting to see if anything else would happen.*

*Shortly after 17:00 local time, just as the sun was setting, a call came in that a sailing vessel's EPIRB (Emergency Position Indicating Radio Deacon) signal had been detected approximately 280 miles east of Hatteras from the Outer Banks. We were told to launch to go out to see if we could find the source of that EPIRB.*

*We did a quick weather check and weather at home station here was very benign, however we knew that a large low pressure system had moved off shore, I checked the weather and still conditions looked to be fairly manageable with winds peak gusts at 30 knots.*

*We quickly got airborne on our C130J and immediately upon going feet wet, I asked my MSO here to bring up our DF430 direction finding capability, and approximately 70 miles off shore or so, we began to get our first bearing point to the source of the EPIRB.*

*This is a significant improvement from my previous experiences in the C130H community, which does not have the advanced systems we do on the C130J. Normally we would have to get much closer to a source, the beacon source in order to get any kind of a hit or composite.*

## Question: And you just said you were 70 miles?

Lt. CDR Driver: *We had a solid block at 70 miles off shore.*

Rear Admiral (Retired) Gilbert by an upgraded
C-130 in Elizabeth City.

### Question: With the C130H systems how close would you have needed to be?

Lt. CDR Driver: *With the H we would have to have gotten much, much closer in order to get any sort of a point at all on the EPIRB signal. I climbed to an altitude, I believe we were up at 230, we had 150 knots of wind on the tail, wind and weather conditions were more severe than forecasted, and as we descended into the area, very close now to the source of the beacon, it was evident that the weather was going to be a lot worse.*

*Winds were sustained at 60 knots on the surface, seas were at 30-40 feet, the ceiling was solid overcast between about 800-900 feet above the water, all the way up to around 12,000-13,000 feet, solid overcast with heavy convective activity of snow squalls, we had icing conditions, sustained moderate turbulence with some severe turbulence and we knew we were going to have our hands full.*

*We had one disadvantage that night, that one, what some people would consider key part of our C4SR system, which was the MMR or multi mode radar, was not functioning. But the depth of the C4SR systems on the airplane made the difference. Under optimal circumstances, we would have had that MMR working, and I probably could have found that sailing vessel even though it bobbed on 30-40 foot seas, I probably could have found it that way, but that system was down.*

*We used the 241 nose mounted weather radar to help us avoid the worst convective activity, weather conditions that could've really damaged the aircraft and been pretty bad for our health and safety. We were able to use that weather radar for that purpose, but we didn't have an MMR to help us find the target sailboat.*

*The good news is, Petty Officer Lee, our MSO was able to take the DF430 equipment, was able to choose lines of bearing on a easterly, westerly heading and on a north-south heading, and we were able to triangulate the source, the signal of that sailing vessel without actually finding it on our radar.*

*So the DF430 paid huge dividends in not only acquiring the signal early, but once we got out there, it helped us pinpoint that location even though we didn't have the benefit of our marine radar.*

*We found the sailing vessel, got over top of it and were able to immediately establish communications with the captain of the vessel, using his VHF-FM, VH-FM band radio channel 16, and we discovered the dire circumstances: this gentleman had been at sea for 5 days at that point, he'd gotten into terrible weather 3 days earlier, he experienced numerous knock downs, the vessel was basically de-masted, his motor was inoperative, he had a prop that was fouled and was essentially dead in the water and could not control his vessel, was getting beamed to the seas and the boat was coming apart underneath of him.*

*We were able to pinpoint his location with the DF430 and then other C4ISR systems like for instance the night vision goggles we wore that night, were absolutely invaluable and critical. The pilots and crew in the back wore NVG's throughout the duration of the mission, we were able – below that overcast, solid overcast layer – to see the vessel.*

*When the gentleman would key his radio, his VH-FM radio and a small light would come on his radio, we could actually see that in the darkness, and we used technique where we just had him take a little hand held penlight, he happened to have a little survival pen flashlight, and he could use that to signal his location.*

*And in those kinds of difficult weather conditions, it was critical using those NVG's to keep our eyes on the target. We also greatly employed our FLIR (Forward Looking Infrared Radar), the FLIR turret, the FLIR ball and camera that night, our MSO was able to keep the FLIR on the target, in fact that system is what provided this picture that you have on the website, and we could maintain much better awareness of the status of the mariner, it was very dark and with the NVG's, we could find the vessel's location, but we couldn't see the fine detail of what he was doing on deck.*

*The FLIR allowed us to know that he was actually still on deck; we didn't need radio transmissions or we didn't need to demand that he use his radio and potentially run down the batteries on his radio. We could see him active on his sailing vessel and we knew that he was still safe and on board. So that helped prevent unnecessary radio communication.*

*We were also able to determine which direction he was steering, to determine the wind and where to actually do our drops. The vessel's aspect is very critical during these drop evolutions, that eventually we've found ourselves performing.*

*And to be able to know where port and starboard was for this captain in relation to our position and sky and the three dimensional world, it was really very helpful because we could literally say to him, turn right, look right off your starboard bow and take a look over here, you will see the airplane coming at you, and he was able to do that and we were able to correlate that in space and time using the FLIR and the aspect ratio that we got from the FLIR.*

*Petty Officer Lee (Missions Systems Officer): I was able to get a couple flares off of that too once we started pointing in the right direction; we talked to him on the 16, and he was able to pop off a couple flares for us.*

Lt. CDR Driver: *We had night vision capability both in the NVG's that we wore on our heads: because those NVG systems were fully integrated into the cockpit design and layout of our instrumentation, there was no conflict between those systems. Also the night vision capability we had in the FLIR as I just demonstrated was pretty critical that night.*

*From a communications standpoint, we relied heavily on really at least 3 radio systems:*

- *Some were very basic: the VH-FM capability talking on channel 16 with the mariner was important*
- *We also were able to utilize our Inmarsat radio, which we call the Steve phone*
- *And Inmarsat communications, that allowed us to call back to Elizabeth City and talk directly to our watch center here and coordinate the follow-on assets, the launch of an additional C130J, as well as coordinate the launch of the H60, that eventually came off the USS Eisenhower and affected the hoist.*

*We were able to coordinate these things through that capability and that was critical. The organic HF baseline radios were critical in maintaining our radio guard and talking to*

RCC (Rescue Coordination Center) Norfolk and getting our tasking and keeping our chain of command informed of what was happening out there on scene. And so those communications were all pretty critical.

So you can see that even though one piece of our C4ISR system, the MMR had failed, we were able to fall back on the other redundant systems and fill that gap and perform the mission.

To make a long story short, we stayed over this gentleman for approximately 4 hours, the vessel continued to get heavily damaged in the seas, and with about 30 minutes left prior to having the H60 from the Eisenhower on scene, we became fuel critical. Just as that was all coming to the head, the vessel was hit by a massive 40-foot wave which caused it to roll 360 degrees.

The mariner, the captain on board was tethered with a safety line to his vessel and he managed to stay with the vessel during that complete roll, when he came back up he was extremely agitated, as you can imagine, and basically said he did not think he was going to be able to last much longer.

Because I was fuel critical, and I had reserved the rescue rafts and the rescue gear in the back for when I needed them, I quickly realized that time was now to launch them. So with the remaining fuel I had on scene, we relied heavily on the computer systems on board the HC130J to set up a rudimentary drop pattern for us.

What we quickly discovered was that because the winds were so strong at 60 knots sustained with occasionally higher gusts, the computer calculated drop solutions that actually became unusable.

If you will, the course that I needed to fly was perpendicular to the wind line; the purpose being we would drop the three rafts into the ocean and the wind would cause the rafts to drift down onto the mariner and his sailing vessel.

Because we were perpendicular to the wind, it required at least 45 degrees of correction in order to keep the plane flying on its required course to make the drop; I had to move 45 degrees.

My heads up display and the entire computer calculated drop data were unusable to me, so I ended up having to revert to a visual only drop maneuver.

Again, the NVG's made that possible, without them I could not have successfully done that, and we were able to perform the drop, push the 3 rafts out just as we were fuel critical, we began to climb to altitude and just before we had to depart we witnessed the rafts drifting down on the position of the sailing vessel.

We communicated two more times with the captain before his radio completely failed. The water intrusion from that roll finally caused his radio to fail, we lost all communications, but we left our final instructions with him in the blind and began to climb out to altitude.

As we were climbing out, the C130J #2 was coming straight in and we passed in flight as he arrived on scene, and I was able to use the on board communication system to pass all the pertinent mission data to the 2nd C130J that was relieving us on scene. And we basically battled those same 150-mile an hour winds back to Elizabeth City and landed with a marginal amount of fuel.

We didn't know the rest of the story until the next day. We found out that just after climbing out and doing a handover with both the C130J that was arriving on scene and the H60 that was now only about 20 minutes out, from the USS Eisenhower, we discovered that just after we left a second 40 foot wave hit the sailing vessel.

This wave completely rolled the vessel again; throwing the master off, his safety line broke so he was a person in the water at that point. He said just as he struggled to the surface to get *his first breath, he looked and very close to him he saw his sailing vessel break apart and sink.*

*So just after we left, he went in the water, he made the statement that he really got a breath, he realized his situation was dire, he made peace with God, but he wasn't ready to give up yet. And he said he got a few breaths, he allowed himself just a little bit of rest from the exhaustion that he was suffering, and he began to swim.*

*Miraculously he said he only took a few strokes swimming and bumped into an unlit object he could not see, it turned out to be the rescue raft, one of the three rescue rafts that we had dropped to him.*

*The raft had been blown upside down in the swells and the sea state, so he did not see the chem. light and strobe that we had attached to it, but he was able to right the raft and climbed into the raft 300 miles offshore, and probably only 15 minutes later the helicopter arrived on scene, saw him in the raft and was able to hoist him to safety out of the raft that we had dropped.*

# THE C4 ISR EFFORT:
# THE VIEW FROM 2010

WE DID several interviews in the period from 2009 through 2011 on building out C4ISR for the USCG. After all, this was the key enabler of building a force that could operate as a security web.

But the most comprehensive interview which we did in this period was one we did in 2010 with Rear Admiral Robert E. Day Jr. at his office.

He assumed the duties as the Assistant Commandant for Command, Control, Communications, & Information Technology and the Director, Coast Guard Cyber Command, Pre-Commissioning Detachment on July 2010. Rear Admiral Day graduated with a Bachelor of Science Degree in Electrical Engineering from the U.S. Coast Guard Academy in 1980. His first tour of duty was in Portland, Maine where he served aboard Cutter Duane as the Damage Control Assistant until April 1982. RDML Day was then assigned to CG Headquarters, Washington D.C. as the Electronics Project Officer for the construction of the 270 foot Medium Endurance Cutters and 110 foot Island Class Patrol Boats.

## The Role of C4ISR in the USCG: Meeting the Challenges for a 21st Century Force

**Question: Could you provide us with an overview of C4ISR works in the USCG?**

Admiral Day: *Let's talk about how C4ISR is used in support of Coast Guard missions. And what changes have occurred, drastic changes, in the last 10 years and the drastic changes that are going to be needed even in the next five. These changes may or may not occur, because they may or may not make the funding threshold. In most cases right now, they are not going to make the funding threshold.*

**Question: C4ISR is essential for a modern Coast Guard to function. Although ethereal to many, the glue, which holds the platforms together, is clearly C4ISR. Could you provide a sense of the shift in performance enabled by the new C4ISR systems?**

Admiral Day: *Let's talk about just the Eastern Pacific drug mission. Let's just use that as an example. In the old days, we literally went down there and bored holes in the water, and if we came across a drug vessel, it was by sheer luck. It might be on a lookout list, and we might happen to see it.*

*Let's fast-forward now to the 2000s and what we've started being able to do. By being able to fuse actionable intelligence, and not only that, but intelligence communicated at light speed. So now, we're to the point where we're telling a Cutter to go point A, pick up smuggler B with load C. And we're doing that in real time with delivery of a common operational picture, which has been fused with intelligence. That was unheard of 10 years ago.*

**Comment: So you're contrasting on the one hand the hunt-and-peck method or the stumble across by chance method, versus having enough information to actually target a problem.**

Admiral Day: *And not only that, taking information from a wide range of intelligence sources and agencies that we can participate with and bringing it in and fusing it. And leveraging all those tools and being able to process that information to figure out anomalies and actually start doing these interdictions.*

**Question: Could you contrast your experiences as a young sailor and a sailor doing the mission now?**

Admiral Day: *Well, it's a whole different framework. The framework is shaped by most of the fusion of the information which is being done off the National Security Cutter. The Cutter is merely a delivery mechanism for capability; the Cutter is now the point of the spear. It has enabled the networks and all the systems back ashore at our Command Centers and our Intelligence Coordination Centers, whether it'd be from Joint Inter-Agency Task Force (JIATF) South or whether it'd be our own, This ability to communicate that to them in real time allows to literally send them a common operational picture: the X is already on your radar screen, and you say go to that target.*

**Comment: So the difference here is that in the first case, you're just throwing a spear out to the ocean.**
Admiral Day: *And hope you hit something.*
**Comment: And where you land, hopefully somebody's near the spear. So the way you're thinking is we**

**have this grid over an area, and your platforms are the customers, so to speak, or the enforcers.**

Admiral Day: *Absolutely. They're the operational element that we are producing information for their mission execution.*

**Question: The C4ISR systems are essential to changing the calculus of operations as well as enabling the USCG in its joint role as well?**

Admiral Day: *Yes. For example, in the eastern Pacific, that's done in JIATF South, which is an interagency task-force, they're doing the lay-down based on the information that they've got. They're getting the intelligence feeds as well we're getting intelligence feed and feeding into it.*

**Comment: I would imagine that the relative "invisibleness" of C4ISR makes support for the systems, notably in funding a continuing challenge? After all, you cannot break a bottle over C4ISR like you do a ship.**

Admiral Day: *The looming budget environment presents significant risks for IT in general. History demonstrates that when budgets get tight, senior management generally starts cutting IT systems, vice other assets, services and personnel. But business research will tell you that this strategy is generally flawed in exactly the wrong approach to take. Because many of the systems that are fielded are being developed to provide exactly the efficiencies and the effectiveness sought in the incremental budget environment.*

*And in no time more so than now, IT provides greater efficiency and effectiveness, whether it be communications amongst varied partners such as you can get, and aggregate response versus stovepipe responses, you get better collaboration and coordination when these systems can talk together. Additionally, you can have that ability to flow information back and forth across lots of boundaries. But killing or significantly reducing an important IT program that is about to be launched negates the prior investment and incurs significant restart expenses.*

*In no time more so than now, IT provides greater efficiency and effectiveness, whether it be communications amongst varied partners such as you can get, and aggregate response versus stovepipe responses, you get better collaboration and coordination when these systems can talk together. Additionally, you can have that ability to flow information back and forth across lots of boundaries. But killing or significantly reducing an important IT program that is about to be launched negates the prior investment and incurs significant restart expenses.*

**Comment: I would assume that there is a challenge of meshing the Department of Homeland Security IT reforms with those of the USCG as well.**

Admiral Day: *Increasing demands for accountability, transparency, standardization, and centralization will require significant investments and/or requirements reductions in Coast Guard IT systems.*

*Let me explain this. Over the last decade, there has been increasing IT regulatory requirements and mandatory program oversight and reporting to our overseers, Congress, OMB and GAO. DOD and DHS have significantly expanded the effort required from Coast Guard personnel and IT systems to produce even the most basic information being mandated.*

*The Coast Guard will be under continuous scrutiny, and in some cases budgetary mandate to collapse existing Coast Guard IT systems into DHS and sometimes and in some cases, government-wide IT systems.*

*The objective of the initiatives to form DHS systems and services is clearly and understandably to shape an IT system that can serve a majority of the needs for all DHS compo-nents. By so doing, it is possible to collapse legacy component systems to achieve standardization and cost savings. The scope of the DHS initiatives is broad. Where possible such common-alities certainly are important and make sense.*

*The objective of the initiatives to form DHS systems and services is clearly and understandably to shape an IT system that can serve a majority of the needs for all DHS compo-nents. By so doing, it is possible to collapse legacy compo-nent systems to achieve standardization and cost savings. The scope of the DHS initiatives is broad. Where possible such commonalities certainly are important and make sense.*

**Question: But I would imagine that the unique USCG mission set requires international partnerships and military relationships and hence IT and C4ISR systems different from much of DHS.  How will this be handled?**

Admiral Day: *The challenge for the USCG is performing its unique mission set mandated by the Congress. And the ques-tion is whether commonality will fit all in order to perform our mission sets.*

*For example, the DHS system may not be able to handle our personnel competencies.  We're military; we have different competencies than a majority of the rest of DHS. We are not solely a law enforcement agency. But again, we're military, and so are our competencies and a lot of our systems that we need to be able to do to track military personnel.*

**Question: And what about interoperability require-ments for working with joint team members not within DHS?**

Admiral Day: *Let's talk about the interagency operations center. We know that particularly in a port, let's look at LA/LB or some of these, Jacksonville, these other locations where we're dealing with the Navy, other federal partners, local partners and state partners all in that port to do port operations.*

*And to work with these folks we need common IT systems. We start using scheduling such that CBP and the Coast Guard are all scheduling off of the same exact system.*

*In the old days, the agencies would go on board ships for inspections sequentially. The Coast Guard would go aboard, they'd do their inspections and ICE would come aboard and they'd do their inspections, and then CBP would come aboard. And so they'd tie the boat up for four to six hours with just these three different visits.*

*Now they all go aboard together and they all execute the mission at the same time. They saved that boat probably two, maybe three hours. And to them, that probably equates to a quarter million dollars, because the longer they keep that boat tied up at that port, means they're not hauling Sonys and stuff to the United States. And so, they're very happy about it. And they don't mind the fact that we come onboard.*

*In the old days, the agencies would go onboard ships for inspections sequentially. The Coast Guard would go aboard, they'd do their inspections and ICE would come aboard and they'd do their inspections, and then CBP would come aboard. And so they'd tie the boat up for four to six hours with just these three different visits.*

*Now they all go aboard together and they all execute the mission at the same time. They saved that boat probably two, maybe three hours. And to them, that probably equates to a quarter million dollars (...) [so] they don't mind the fact that we come onboard.*

*But it takes the IT systems for these guys to sit down*

*jointly to decide which ships they're going to target, which ones are going to go aboard, who's going to do it. With these scheduling systems and with the ability to see across each other's systems and to shape commonality, we can work together to execute a simultaneous inspection. This all fits into a port interagency operations center, and some of the systems that we're trying to build, so that that we can share data.*

**Question: So joint decision-making is crucial in the port enabled by IT systems that work for the civil, law enforcement and military authorities?**

Admiral Day: *That's exactly right. But it is not just IT systems; it is building an effective process that uses the IT systems. My IT system is not going to be effective, it doesn't matter what system I develop, it's not going to be effective unless you change your processes. Or refine your processes. And this is the piece that does not show up easily in flow charts or funding efforts.*

**Question: So support for decision-making tools and decision-making processes is essential to C4ISR success?**

Admiral Day: *For example, the USCG is building such a system at Portsmouth called Watchkeeper. This is a scheduling interface, but this is essentially a decision making tool, and a situational awareness tool particularly for the sector level.*

*But the funding for this program is on life support. But this is the exact tool that we need. What we're doing is building an entire architecture, so Watchkeeper is the level that's going to be at the sector command center level.*

*C3CEN is CG center of excellence for these systems; they*

*fuse in a whole bunch of other capabilities. They fuse in NAIS (National Automatic Identification System). They do radars, etc. So, they are the sensor people as well, so they've got the fusion: they build the fusion system for the sensors that they're placing out there. They're also DGPS (Differential Global Positioning System).*

*C3CEN is CG center of excellence for these systems; they fuse in a whole bunch of other capabilities. They fuse in NAIS (National Automatic Identification System). They do radars, etc. So, they are the sensor people as well, so they've got the fusion and they build the fusion system for the sensors that they're placing out there. They're also DGPS (Differential Global Positioning System).*

*So, it's a family of systems and this is where we're trying to move to, even in the deepwater environment, because now we got the pallets aboard the MPA (Maritime Patrol Aircraft); we got the systems aboard the C-130Js. And I'm not going to ever use that term of systems of systems. But we're building this architecture so that they all plug into a common enterprise bus.*

**Question: Could you give us a sense of how commonality with some of the DHS elements is being enhanced?**

Admiral Day: *I do a lot with Customs and Border Patrol (CBP). Right now, most of my HF communications, which primarily service my aircraft, particularly helicopters, C-130s, and sectors to a certain extent is done using the customs over the horizon enforcement network, COTHEN. Because I have a significant HF infrastructure, and they have a lot as well, why not organize differently?*

*If we link our stuff together, in other words, I use some of my HF transmitters, and they use some of theirs, we now have a global grid that we can use versus us with our little patch-work pieces everywhere.*

*So now, some of my high frequency radio communications are being done on in cooperation with CBP to the point where is it's all run out of one operations center down in Florida that there is Coast Guard watch standers sitting in CBP facility doing this joint piece for HF radio networks.*

*There's even more that could be said if the investments were done right that we could leverage Rescue 21 for some of these capabilities.* *

## Question: And what is the impact on leveraging commercial systems to solve some of the C4ISR problems?

Admiral Day: *Let me just talk about Haiti. What did they do? They took and they put up a 3 G cell site, and they handed everybody out cell phones and they leveraged the heck out of them. And not only that, they leveraged things like text; they leveraged video back and forth across them. And it came down to a common standard that everybody knew how to use, because they all know how to use a dial tone, there is no training necessary.*

*And it even gets to the point of tactical communications when I move out. If I turn up a 3G cell site on my cutter, on my national security cutter, and I've got 3G communications capability or 4G by then, it's a small handheld device that I can hand to any 18-year-old, and they know automatically how to handle it.*

*Not only that, they'd hold it right here and they turn the video on, and now I got a live video feed from my boarding*

---

* Rescue-21 is a coastal distress and safety system that also supports Coast Guard C4ISR.

*team. And if I want to, I can encrypt it such that they can't, you know, I'll just go ahead and put AES encryption into it, because I'm running my own private cell company. Or if I want to, I can open it up to somebody else saying what's your mobile number? Okay. You're a part of my team.*

*And it's going to get boiled down to something like that. Same thing at Deepwater Horizon, because nobody has the same radio systems: BP had a private radio system, all the counties had private radio systems, the parishes had private radio systems; it's no different than the fire department , which can't talk to the ambulances, which can't talk to the police department. They all go back to the least common denominator that they know. They bring out cell phones.*

**Question: You mentioned earlier that you're using your 3G private or your 4G network, but you're creating all these private cell phone company: does that mean that you are essentially leveraging the network?**

Admiral Day: *I'm just leveraging the network. We did it in New York during 9/11, because we had lost it when the second tower went down—when the south tower went down, we lost everything. We were blind, because all of our telecommunication circuits were in that south tower. So that means the VTS lost all of their radar signals, they lost all the VHF communications, we were deaf, dumb and blind, because not only that, we lost our telephones.*

**Question: But what do you mean by the phrase, you created your own cell phone network?**

Admiral Day: *What I'm saying is, is that at a moment's notice I can turn up my own infrastructure that replicates commercial, off the shelf, no different than a Sprint or an AT&T. But for my contingency or even my day-to-day operations, I can run a network that I can control, I can let in who I want in, but also interfaces cleanly with the public.*

**Question: So as long as you have some spectrum, and your cell phones wirelessly linked to your network, you can do it. And your point is that no one needs training for that?**

Admiral Day: *Indeed, they know how to use the device. I haven't seen a contingency response yet that one could say one radio system satisfied anything.*

**Comment: But you lose that capability when you go well offshore. In most places, you don't have affordable connectivity. Worldwide commercial satellite systems are expensive and not subsidized by millions of users**.

Admiral Day: *That is definitely true when we have an intense operation well offshore. Two things are helping here; many intense incidents are in the coastal regions where there is relatively good connectivity. Satellite systems are becoming more available although many are expensive. Again, the commercial world with cruise ships and other vessels needing lots of bandwidth are causing the market place to respond with dramatically increased capability.*

*One example was the Deepwater Horizon response. Because a morale issue aboard the rigs was that when they're out there for four weeks, they couldn't call home. So they worked with the cell phone companies, and the cell phone companies said hey, I'll put a cell site out there for you.*

*And by the way, all those shrimpers and all the fishermen who were within 50 miles of there, I'm going to get them too. So they had a profit motive for doing so. We were able to tap into this commercially provided capability. Similar programs are happening in the more offshore areas. Again, this re-emphasizes the critical role our Coast Guard plays internationally with organizations such as the International Maritime Organization. They create regulations for international shipping for use of radio communications and electronic systems in general for maritime safety, security and environmental protection. These programs lay the foundations for interoperability and enable commercial vessels to communicate among themselves and us. Absent some organization to enable all this, we would have towers of babble, especially in emergencies.*

*And let me just give you one of the problem sets that Deepwater Horizon presented. We flew almost 2,000 people in there very quickly. And one of the things that we learned from Katrina that a major problem was keeping track of everybody was a very difficult thing. And we ended up with people who were down there for two months, and where have you been? What were you doing? And they came to me and said we got to do a better job of tracking personnel.*

*Well I said well give them a cell phone, give them one with GPS on it, keep track of their cell phone number and literally using Google Earth and the capabilities that are available, we know where everybody is. I can track personnel down to about 10 meters if it's enabled. I said just give them the phones, turn it on, we'll use the commercially available services and we'll track the sons of a gun.*

**Question: To wrap up, how would summarize the contribution of C4ISR to USCG missions and capabilities?**

Admiral Day: *The American people's investment in our new capital platforms that they are graciously delivering to us will be sub-optimized if we cannot do the rest of the systems piece and link all this stuff together.*

*With the new C4ISR systems, I can fuse the data. And such capabilities are the lynchpin that enhances the mission execution from a limited set of platforms, and improves their efficiency and their effectiveness to get the job done. That's why investment in systems such as Watchkeeper, Coastwatch or the sensor systems which are going to go aboard the cutters, and the pallets aboard the CASAs and the 130Js, are crucial. With such tools, we can shape more effective decision-making and leverage our always limited numbers of platforms.*

By Robbin Laird and Ed Gilbert
December 9, 2020

# PART 6

# THE USCG AND CRISIS RESPONSE

Part six of "Always Ready, Persistently Under-Resourced" focuses on the United States Coast Guard's pivotal role in crisis response through two detailed case studies. This part serves as the empirical core of the book's larger argument, illustrating both the strategic value and vulnerabilities of the Coast Guard in real operational terms.

Together, these studies underscore recurring tensions in Coast Guard missions: balancing global responsibilities with domestic priorities, meeting operational demands amidst political and resource limitations, and sustaining a culture of excellence under chronic underfunding.

These cases diagnose not just what the Coast Guard achieves when equipped and empowered, but also the fragility of those successes given aging platforms, workload strain, and insufficient resourcing.

Part Six hus functions as both illustration and critical diagnosis, making clear that the Coast Guard's enduring readiness and indispensability depend on addressing persistent underinvestment and sustaining its modernization efforts.

# 1

# THE COAST GUARD'S
# KATRINA RESPONSE

WHEN HURRICANE KATRINA made landfall on the Gulf Coast on August 29, 2005, it became one of the most devastating natural disasters in American history.

More than 1,300 people lost their lives, damage stretched over a 90,000 square mile area, and over a million people were driven from their homes.[*]

Amid this catastrophe, the United States Coast Guard emerged as the standout federal responder, executing what would become one of the largest rescue operations in the service's history. Yet the crisis also exposed fundamental structural limitations that continue to constrain the service two decades later.

The Coast Guard's performance during Katrina stood in stark contrast to the widely criticized federal response. Of the estimated 60,000 people who needed to be rescued from rooftops and flooded homes, Coast Guardsmen saved more than 33,500, including

---

[*] A major source for the data in this chapter is taken from the following: https://www.gao.gov/products/gao-06-903

rescuing from peril 24,135 lives and evacuating 9,409 medical patients to safety.

At the peak of operations, the Coast Guard was rescuing 100 people per hour by air and 750 people per hour by boat. Over 5,600 Coast Guard personnel participated in the response efforts, with more than 3,400 serving in the region on a single day, representing over ten percent of the entire service.

Forty percent of the Coast Guard's helicopters were deployed to the Gulf Coast, with aircraft serving from every Coast Guard air station from Barbers Point, Hawaii to Kodiak, Alaska to Cape Cod, Massachusetts.

The magnitude of this achievement becomes even more impressive when one considers the operating environment. Coast Guard units conducted rescue operations while New Orleans was without power, potable water, or functioning sewage systems. They faced resistance from some local law enforcement officials and the constant threat of armed gangs roaming the streets.

Communications infrastructure was destroyed, with radio towers toppled and phone lines severed, forcing Coast Guard personnel and auxiliarists to source satellite phones and portable radios to maintain minimal operational capability.*

## Admiral Thad Allen and
## Federal Leadership

The Coast Guard's organizational success was personified by Vice Admiral Thad Allen, who was called on Labor Day, September 5, 2005, one week after Katrina's landfall, by Secretary of Homeland Security Michael Chertoff at President George W. Bush's direction.

Allen was asked to assess the situation in New Orleans and take over federal relief efforts following the removal of FEMA Director

---

* https://www.history.uscg.mil/Our-Collections/Documents-Publications/igphoto/ 2001835784/

Michael Brown. On September 9, Secretary Chertoff formally appointed Allen as Principal Federal Official for Hurricane Katrina response operations throughout the Gulf Coast region.

Allen brought to the role both operational expertise and a leadership style forged through decades of Coast Guard service. The son of a retired Coast Guard chief petty officer who had joined during the Great Depression, Allen possessed an ability to effortlessly mask a keen intellect with a seaman's salty humor.

His approach was notably different from that of Lt. General Russell Honoré, the colorful, cigar-chomping Army officer heading military operations. Allen's understated style and steady temperament proved essential in navigating the political maelstrom surrounding the disaster.

Allen's first major insight came within 24 hours of arriving on scene. Flying over New Orleans on September 6, he landed at a makeshift zone near the Convention Center and reached a crucial conclusion: for a week, responders had been "trying to solve the wrong problem."*

Resources were pouring into the region following President Bush's emergency declarations, but the City of New Orleans had lost continuity of government and couldn't manage the resources being provided. Someone needed to take tactical operational control to organize the relief mission, a role the Coast Guard was uniquely equipped to fulfill.

## The Organizational Factors Behind Coast Guard Success

The Government Accountability Office, in its comprehensive review of the Coast Guard's Katrina response, identified several key factors that enabled the service's effectiveness. These factors provide insight

---

* https://nationalcoastguardmuseum.org/hurricane-katrina/timeline/miracles/

into why the Coast Guard succeeded where other federal agencies struggled.

- First, the Coast Guard's operational principles promoting leadership, accountability, and decentralized decision-making enabled personnel to take responsibility and action based on relevant authorities and guidance. This stood in sharp contrast to other federal agencies that were paralyzed waiting for formal orders and coordination structures. Coast Guard units self-deployed and began rescue operations immediately, demonstrating the service's culture of initiative.
- Second, the Coast Guard relied on standardized operations and maintenance practices that allowed for greater flexibility in deploying personnel and assets from any operational unit. This interoperability meant that Coast Guard forces from across the country could seamlessly integrate into Gulf Coast operations without extensive coordination or retraining. The service operates under the principle that any qualified Coast Guard member should be able to step into a role anywhere in the service, a capability that proved invaluable during the massive personnel surge Katrina required.
- Third, the Coast Guard had up-to-date and regularly exercised hurricane plans. Having prepared for and responded to five named storms earlier in 2005, Coast Guard personnel in the Gulf region were well-versed in hurricane response procedures. Admiral Robert Duncan, Commander of the Eighth Coast Guard District, had authorized the evacuation of dependents and relocated elements of the District Staff from New Orleans prior to Katrina's landfall, implementing a strategy of first preserving Coast Guard personnel and resources so they could then respond effectively after the storm. The

service evacuated 18 small boat stations to predetermined inland military installations precisely so they would not be trapped in flooded coastal stations and unable to respond.

- Fourth, the Coast Guard's daily mission set prepared it for disaster response in ways that other agencies could not match. Search and rescue, marine environmental protection, and management of maritime commerce are missions the Coast Guard performs every day. While the scale of Katrina was unprecedented, the fundamental tasks were familiar. As one Coast Guard officer testified to Congress: "These are missions that we do every single day, search and rescue, response to collisions, response to oil spills. I maintain a 24-hour, 7-day-a-week watch."*

- Finally, the Coast Guard's experience operating in both military and civilian contexts proved essential. As a military service that routinely operates in civilian environments, the Coast Guard bridged the gap between military efficiency and civilian sensitivity in ways the Department of Defense struggled to achieve. Coast Guard personnel are trained to work hand-in-glove with local and state officials, as well as with multiple federal agencies. This cultural attribute enabled effective coordination when other agencies were mired in bureaucratic conflicts over jurisdiction and authority.

## Operational Innovation and Adaptation

Despite extensive preparation, the Coast Guard faced several unique challenges during Katrina that required rapid innovation. The service's standard search and rescue tactics were suddenly inade-

---

* https://www.congress.gov/event/109th-congress/senate-event/LC12130/text

quate as rescue teams faced debris-filled floodwaters and urban rooftop rescues unlike anything they had trained for.

Captain David Callahan, commanding officer of Aviation Training Center Mobile, recalled the need for immediate tactical adaptation:

> We're going to have to change our tactics. And I'm not sure who it was, I think it was the XO [Executive Officer], ordered folks to go out to Home Depot and that night we bought every wood ax and saw we could find and started outfitting our rescue swimmers with those to adapt to this new urban rescue environment that we were in.*

This improvisation which Callahan jokingly worried might result in him "going to jail for buying all those axes and saws" exemplified the Coast Guard's adaptive problem-solving culture.

Aviation Training Center Mobile became the largest air station in Coast Guard history, operating over 43 aircraft, flying 1,193 sorties and 2,202 flight hours. Air Station Houston aircraft flew more than 164 flight hours in 106 sorties, saving 691 lives. Air Station Clearwater saved 1,165 lives and deployed 708,000 pounds in 236 "sling loads," including dumping sandbags on breached levees. Coast Guard C-130 aircraft served multiple roles beyond traditional search and rescue, functioning as communications platforms, conducting surveillance missions, transporting personnel and supplies, and even evacuating stretcher patients to receiving hospitals.

## The Resource Constraints Katrina Exposed

While the Coast Guard's performance was lauded, the crisis brutally

---

* https://www.mycg.uscg.mil/News/Article/4293466/learning-from-disaster-how-katrina-helped-us-prepare-for-future-catastrophes/

exposed the service's structural limitations. These constraints stemmed from a fundamental mismatch between expanding mission requirements and available resources, a tension that has only intensified in the two decades since Katrina.

The Coast Guard is responsible for conducting eleven statutory missions spanning homeland security and non-homeland security functions. These include search and rescue, marine safety, aids to navigation, living marine resources (fisheries law enforcement), marine environmental protection, ice operations, ports and waterways security, drug interdiction, migrant interdiction, defense readiness, and other law enforcement. This breadth is unmatched by any other military service, yet the Coast Guard is the second smallest branch of the U.S. military in terms of personnel.

The chronic funding challenge became evident during Katrina. The service had to cannibalize resources from other mission areas to concentrate forces in the Gulf. Drug interdiction, fisheries enforcement, port security in other regions, and other missions were degraded as assets were pulled to support Katrina operations. Unlike the Navy or Air Force, which maintain some strategic reserve capacity, the Coast Guard operates at near 100 percent capacity during normal operations. There was no reserve force to draw upon, the service had to strip other operational areas to generate the surge capacity Katrina required.

## Conclusion: Culture Cannot Indefinitely Compensate for Capacity

Hurricane Katrina revealed the United States Coast Guard at both its finest and most vulnerable. The service executed one of the most successful large-scale rescue operations in American history, demonstrating operational excellence, adaptive leadership, and organizational agility that other federal agencies could not match. Admiral Allen's leadership helped salvage a federal response that had veered toward comprehensive failure, and Coast Guard rescue swimmers, pilots, boat crews, and support personnel performed with courage and professionalism under extraordinarily difficult conditions.

Yet Katrina also exposed uncomfortable truths about structural constraints. The Coast Guard's success came from organizational culture, decentralized leadership philosophy, and individual heroism, not from adequate investment in capabilities or sustainable resource allocation. The service achieved extraordinary results by operating at an unsustainable tempo, stripping other mission areas of coverage, and exhausting its personnel.

Two decades after Katrina, the fundamental challenge persists: the Coast Guard faces expanding mission requirements without proportional resource increases. The service has improved some capabilities, strengthened acquisition management, and incorporated lessons learned from Katrina and subsequent crises. However, questions remain about how long exceptional culture can compensate for structural underfunding and whether "doing more with less" has reached the point of becoming "doing less with less."

The Coast Guard proved during Katrina that it is capable of rising to meet extraordinary challenges. The question for policymakers is whether America's smallest armed service will have the resources to do so the next time the nation calls.

As Admiral Allen himself observed during a later interview, Katrina was not merely "a once in a career event" or even "a once in a lifetime event" but it was "a once in the services' lifetime event."

The Coast Guard met that test, but the service's capacity to meet the next one remains uncertain without addressing the persistent mismatch between missions and resources.

# 2

# THE DEEPWATER HORIZON CRISIS

WHEN THE DEEPWATER HORIZON mobile offshore drilling unit exploded on the evening of April 20, 2010, approximately 45 miles southeast of Venice, Louisiana, it triggered what would become the largest marine oil spill in American history.[*]

The explosion resulted in the deaths of 11 workers and initiated a cascade of events that would test the United States Coast Guard to its absolute limits.

What emerged from this crisis was not only the largest environmental disaster response in the service's history but also a stark revelation of the resource constraints and capability gaps that had developed within one of America's most essential maritime services.

## The Scale of the Challenge

The magnitude of the Deepwater Horizon disaster was unprecedented. An estimated 60,000 barrels of oil gushed into the Gulf of Mexico each day for 87 days, ultimately releasing over 200 million

---

[*]   https://nationalcoastguardmuseum.org/articles/deepwater-horizon/

gallons of crude oil into U.S. waters.* The oil slick covered thousands of square miles of water and affected nearly 4,500 miles of Gulf shoreline.

As Captain Duke Walker, Federal On-Scene Coordinator for the U.S. Coast Guard, would later observe: "Even Exxon Valdez pales in comparison to the volume, scale, number of Coast Guard resources, how much time has been devoted to [Deepwater Horizon], all far exceed any previous event."[†]

The Coast Guard's response began immediately after the explosion at approximately 10:00 p.m. on April 20, when District 8 Command Centers were notified of the fire. The service initiated search and rescue efforts, coordinated firefighting, and established an incident command post. At the peak of response operations during the summer of 2010, there were over 47,000 response personnel involved, with nearly 9,000 from the USCG alone. This massive deployment would continue for years, with the Coast Guard leading cleanup and evaluation efforts along Gulf shoreline segments from April 2010 through April 2014.

## Leadership and Command Structure

On April 30, 2010, Homeland Security Secretary Janet Napolitano announced that Admiral Thad W. Allen, the recently retired Commandant of the Coast Guard, would serve as the National Incident Commander for the federal government's response to the spill. This designation was historic, the first time a spill had been declared one of "national significance" under provisions established after the 1989 Exxon Valdez disaster. The declaration triggered a national command structure designed to coordinate efforts at all levels, from

---

* https://www.defensemedianetwork.com/stories/deepwater-horizon-bp-oil-spill-coast-guard-response-then-and-now/.

† https://www.slideserve.com/guri/u-s-coast-guard-deepwater-horizon-incident-response-summary#google_vignette; https://nationalcoastguardmuseum.org/articles/deepwater-horizon/

local to federal, and to "de-conflict problem areas" across multiple agencies and jurisdictions.

Admiral Allen's appointment symbolized both the Coast Guard's central role and the unprecedented nature of the challenge. As National Incident Commander, he worked closely with the Environmental Protection Agency, Department of Homeland Security, the Departments of Defense, Interior, Commerce, and Health and Human Services, state and local entities, and BP to bring unity to response operations.* The Coast Guard's role as Federal On-Scene Coordinator meant the service was responsible for directing all response efforts to contain and clean up the oil spill under the National Contingency Plan.

## Resource Limitations:
## Equipment and Personnel Gaps

The Deepwater Horizon response exposed critical resource shortages within the Coast Guard. While the service successfully coordinated the massive multi-agency response, demonstrating strong command and control capabilities, it faced severe equipment deficiencies that hampered operational effectiveness.

The Coast Guard lacked sufficient specialized pollution response vessels, skimmers, and containment booms needed for a disaster of this magnitude. The service was forced to rely heavily on private contractors, borrowed assets from other agencies, and international partners to mount an adequate response.† This equipment gap reflected years of chronic underfunding and deferred maintenance across the Coast Guard's aging fleet.

The personnel strain was equally significant. Over 8,500 Coast Guard personnel were deployed as responders to the oil spill. These

---

* https://www.idb.org/our-team/db-executive-fellows/thad-w-allen/
† Based on analysis of Coast Guard resource deployment patterns and reliance on contractor support during the response.

deployments created severe operational stress, as Coast Guard units had to maintain their other critical missions, search and rescue, port security, drug interdiction, and fisheries enforcement while dedicating substantial resources to the oil spill response for months and even years. The sustained nature of the operation, which lasted from April 2010 through 2015 when operational oversight was finally transferred to local Coast Guard field commanders, demonstrated the service's commitment but also revealed how thin its resources were stretched.

The health consequences for Coast Guard responders further illustrated the human cost of the response effort. Studies of the Deepwater Horizon Oil Spill Coast Guard Cohort found that responders had elevated risks for cardiovascular conditions, respiratory symptoms, and other health issues associated with crude oil and dispersant exposures. More than half of Coast Guard responders (54.6%) were exposed to crude oil during deployment, and nearly one-fourth (22.0%) were exposed to oil dispersants. These exposures resulted in acute respiratory symptoms, with coughing being the most prevalent (19.4%), followed by shortness of breath (5.5%) and wheezing (3.6%).[*]

## Technical Capability Gaps

Beyond equipment shortages, the crisis highlighted significant capability gaps in the Coast Guard's technical expertise. While the USCG maintained regulatory authority over offshore drilling operations, it had limited technical expertise in deepwater drilling operations at extreme depths compared to industry specialists. The Macondo well blowout occurred at approximately 5,000 feet below sea level, presenting unprecedented technical challenges for containment efforts.[†] This depth exceeded the experience and equipment

---

[*]   https://pmc.ncbi.nlm.nih.gov/articles/PMC5811337/
[†]   https://blog.response.restoration.noaa.gov/early-days-and-hours-deepwater-horizon

capabilities of most government responders, creating challenges in making rapid technical assessments of containment options.

The Coast Guard's regulatory oversight role had also been undermined by broader structural issues. Between 2000 and 2010, the Coast Guard had issued pollution citations to Deepwater Horizon 18 times and had investigated 16 fires on the platform.[*] However, these citations had failed to prevent the disaster, raising questions about the effectiveness of the Coast Guard's inspection and enforcement regime in an era of increasingly complex offshore drilling technology.

The joint investigation conducted by the U.S. Coast Guard and the Bureau of Ocean Energy Management, Regulation and Enforcement revealed numerous systems deficiencies and acts and omissions by Transocean and its Deepwater Horizon crew that had an adverse impact on the ability to prevent or limit the magnitude of the disaster. While these findings primarily implicated the private operators, they also highlighted the Coast Guard's limited capacity to provide real-time oversight of such technologically advanced operations.

## The Deepwater Program: Context of Chronic Underfunding

The resource limitations exposed by the Deepwater Horizon crisis were not accidental but rather the result of decades of constrained budgets and deferred recapitalization. The Coast Guard's Deepwater Program, initiated in the 1990s to modernize and replace the service's aging fleet of ships and aircraft, had been plagued by cost overruns, schedule delays, and funding shortfalls.[†]

The Deepwater Program was originally estimated at $17 billion when the contract was awarded in 2002, but by 2006, the cost had risen to $24 billion. However, even this expanded budget proved inadequate. By 2011, the Coast Guard had developed baselines for

---

[*]   https://en.wikipedia.org/wiki/Deepwater_Horizon_explosion
[†]   https://www.gao.gov/products/gao-11-743

some assets that indicated the estimated total acquisition cost could be as much as $29.3 billion, or about $5 billion over the approved baseline. Coast Guard and Department of Homeland Security officials acknowledged that the annual funding needed to support all approved Deepwater acquisition program baselines exceeded current and expected funding levels.

This chronic underfunding created what Coast Guard officials described as a "readiness gap." Legacy cutters were operating free of major equipment casualties less than 50 percent of the time, despite investment per operational day increasing by over 50 percent over six years. This declining readiness directly impacted the quantity and quality of Coast Guard "presence", the service's ability to accomplish all missions simultaneously.

The Deepwater Program's struggles exemplified the broader problem of maintaining a multi-mission maritime service within constrained budgets. As one 2005 congressional hearing noted, "the Coast Guard is experiencing a continuing decline in fleet readiness," with the resulting readiness gap "negatively impact[ing] both the quantity and quality of Coast Guard 'presence', critical to our ability to accomplish all missions."* When the Deepwater Horizon crisis struck in 2010, these structural weaknesses were fully exposed.

## Lessons Applied and Reforms Implemented

Despite its resource limitations, the Coast Guard's organizational response to the Deepwater Horizon crisis was remarkably effective. As the Coast Guard stated in its 2015 assessment on the fifth anniversary of the incident: "Organizationally, the Deepwater Horizon response was the most effective ever seen for a spill of its magnitude." However, the Coast Guard also acknowledged a

---

* https://www.govinfo.gov/content/pkg/CHRG-109shrg62473/html/CHRG-109shrg62473.htm

sobering reality: "Only a small percentage of the oil was recovered, however, which emphasizes the importance of prevention."[*]

In the aftermath of the crisis, the Coast Guard implemented numerous reforms to strengthen its oil spill response capabilities. These included enhancing the Response Resource Inventory System, the database listing all pollution response equipment in the United States, to better prepare for emerging threats. The service also institutionalized common operational picture technologies, promoting tools like NOAA's Environmental Response Management Application (ERMA) to support operational decision-making by Federal On-Scene Coordinators and stakeholders.

The Coast Guard recognized that new challenges were emerging even as it absorbed lessons from Deepwater Horizon. As the service noted in 2015, "offshore and inland domestic oil and gas production has reached new levels and operations in the Arctic are predicted to continue to grow over the next several decades." These evolving mission requirements would demand continued investment in capabilities and equipment.

## Strategic Implications: The Case for Recapitalization

The Deepwater Horizon crisis ultimately reinforced the arguments for Coast Guard recapitalization and highlighted the risks of chronic underfunding for a service with expanding responsibilities. The Government Accountability Office had warned before the spill that the Deepwater Program baseline was no longer achievable and recommended a comprehensive study to assess the mix of assets needed in a cost-constrained environment. The crisis validated these concerns, demonstrating that inadequate investment in moderniza-

---

[*] https://www.dhs.gov/news/2015/04/22/uscg-statement-record-house-committee-natural-resources-hearing-titled-innovations

tion directly compromised the Coast Guard's ability to respond to major disasters.[*]

Admiral Allen's leadership during the crisis exemplified the Coast Guard's institutional competence in command and control, interagency coordination, and adaptive problem-solving. His ability to create what he called a "unity of effort" among federal agencies, state governments, BP, and thousands of responders demonstrated the service's organizational strengths. However, even the most effective leadership cannot fully compensate for inadequate resources and aging equipment.

The health consequences for Coast Guard responders, documented in subsequent studies showing elevated risks for cardiovascular and respiratory conditions years after their deployments, underscore the human cost of resource constraints. When personnel are stretched thin and deployments are extended due to insufficient equipment and manpower, the service members themselves bear the burden.

## Conclusion

The Deepwater Horizon crisis revealed the U.S. Coast Guard as an organization of exceptional competence operating under severe resource constraints. The service successfully coordinated the largest environmental disaster response in its history, demonstrating remarkable command and control capabilities, interagency coordination skills, and adaptive leadership.

The crisis exposed critical gaps: insufficient specialized pollution response equipment, limited technical expertise in deepwater drilling operations, personnel stretched across multiple simultaneous missions, and an aging fleet compromised by years of deferred recapitalization. These shortcomings were not failures of Coast Guard

---

[*]  https://media.defense.gov/2024/Jun/21/2003489705/-1/-1/0/2009_GAO_DEEPWATER_09620T.PDF

leadership or personnel but rather the predictable consequences of chronic underfunding and inadequate investment in modernization.

As the Coast Guard continues to face evolving maritime security challenges, from Arctic operations to increasingly sophisticated offshore drilling, the lessons of Deepwater Horizon remain relevant. The service demonstrated that organizational excellence and adaptive leadership can achieve remarkable results even with limited resources.

However, the crisis also made clear that continued expansion of Coast Guard missions without commensurate investment in equipment, personnel, and capabilities creates unacceptable risks, to the environment, to the nation's maritime security, and to the Coast Guard personnel who serve on the front lines of every crisis.

The Deepwater Horizon response showcased both the Coast Guard's indispensable role in national disaster response and the urgent need for the sustained investment required to maintain that capability. Future crises will test the service again, and the question remains whether policymakers will provide the resources necessary to match the expanding missions they assign to this essential maritime service.

# The Deepwater Drilling Challenge:
# Moving Beyond Blame to
# Build Effective Stewardship

August 8, 2010

By Robbin Laird and Rear Admiral Ed Gilbert, (Retired)

The tragedy of the Gulf oil spill threatens to devolve into a predictable blame game, a familiar pattern that magnifies disasters rather than mitigates them. This need not be so. Crises of this magnitude provide rare opportunities to assert genuine leadership and shape effective approaches for future challenges. The question is whether we will seize this moment or squander it in recrimination.

The fundamental challenge before us is clear: establishing proper government stewardship for the deepwater drilling enterprise and fundamentally reshaping the public-private partnership governing such activities. Make no mistake, we and other nations will continue drilling in deepwater environments. The global energy equation demands it. Simply planning for a green future, however desirable, remains an exercise in aspirational politics rather than realistic government adaptation to meet near-term energy needs. The transition to alternative energy will take decades, and during that time, deepwater drilling will remain a critical component of our energy infrastructure.

The current regulatory framework has proven inadequate to the task. To execute a proper stewardship function, the US government team needs comprehensive reshaping, substantial recapitalization, and strategic strengthening across multiple dimensions. This is not merely about incremental improvements but about fundamental transformation of how we approach offshore energy development.

- First, we face a structural reshaping challenge. To date, the Coast Guard has maintained responsibility for inspections above the waterline, while the Minerals Management Service handles operations below the surface. This artificial division creates dangerous gaps in oversight and accountability. We need a unified inspection regime with shared powers that can oversee the entire enterprise seamlessly, from surface operations to seafloor installations. The current bifurcation of responsibility allows critical safety issues to fall through bureaucratic cracks.
- Second, and perhaps most critically, government presently lacks the scientific and technological expertise necessary to oversee the deepwater drilling enterprise

effectively. The technology has advanced far beyond the government's capacity to monitor and regulate it. We must recruit and retain world-class scientific and technological experts to serve as genuine partners for the Coast Guard and MMS in ensuring effective oversight. This requires creative approaches to compensation and career development that can compete with private sector opportunities. Without this expertise, government inspectors become dependent on industry assurances rather than independent verification.

- Third, we must acknowledge a fundamental reality: the US government will never possess all the expertise required for comprehensive scientific and technological evaluation of deepwater operations. This capability cannot be fully insourced. Therefore, developing an effective scientific and technological partnership with responsible elements of the private sector becomes essential for success. This partnership must be structured to maintain government independence and authority while leveraging private sector innovation and expertise. The challenge is creating collaborative relationships that enhance rather than compromise regulatory integrity.

- Fourth, the Coast Guard and other inspection and oversight agencies require full resourcing to meet this expanded challenge. The current trajectory of budget cuts to the Coast Guard represents precisely the wrong approach at the wrong time. New funding should be invested in building the Coast Guard's role and expertise in deepwater oversight. The service currently lacks even basic tools for situational awareness and management below the waterline. Additional qualified inspectors must be hired rather than proceeding with the planned personnel reductions in Coast Guard staff. Effective regulation requires resources commensurate with the risks being managed.

- Fifth, the Coast Guard needs fully funded rapid response task forces specifically designated for crises of this nature. These capabilities should be built around additional assets such as the new national security cutters, creating dedicated teams available for immediate intervention when disasters strike. Crisis response cannot be treated as an extension of normal operations. It requires surge capacity that can be deployed without compromising routine missions. The current approach of robbing Peter to pay Paul leaves critical gaps in coverage.

- Sixth, and most troubling, the Coast Guard has been forced to redeploy many of its assets from other regions to address the Gulf crisis, leaving the nation vulnerable elsewhere. This is fundamentally unacceptable and can only be solved by accelerating the overall recapitalization of the Coast Guard's rapidly depleting core capabilities. The service cannot continue to make impossible choices between competing critical missions. The fleet modernization program must proceed at full speed to ensure adequate presence across all mission areas.

The Administration has articulated its commitment to effective participation in protecting the global commons. No clearer example of exercising leadership in the global commons exists than shaping an effective response to the challenges of deepwater drilling. The Gulf oil spill is not simply an American affair confined to regional waters. Ocean currents will carry consequences throughout the Atlantic and beyond, making this a truly international concern that demands American leadership.

What we need now are substantive hearings focused on how to properly move forward, hearings that result in concrete recapitalization and stewardship budgets rather than political theater. We have already witnessed sufficient inquiries into the failures of the oil industry. It is time to focus in a serious, no-nonsense way on defining the proper U.S. government role in overseeing deepwater operations. This requires moving beyond blame to build institutional capacity.

The path forward demands both immediate action and long-term commitment. We must invest in expertise, modernize oversight frameworks, build dedicated response capabilities, and resource these efforts adequately. The American people and the international community should expect no less. The stakes are too high, and the consequences of continued inadequate oversight too severe, to accept anything short of fundamental transformation in how we approach deepwater drilling stewardship.

The time for half-measures and business as usual has passed. The question is whether we have the political will to build the capabilities that effective stewardship demands.

# PART 7

---

# FAST FORWARDING

Part seven of the book is where the Coast Guard's recent past collides with the emerging future, and the core themes of this book, modernization, mission growth, and chronic under-resourcing, come into even sharper relief.

In the earlier chapters, we followed the service through the Deepwater era, the post-9/11 homeland security turn, Katrina and Deepwater Horizon, the fielding of the National Security Cutter, and the first wave of C4ISR upgrades that began to turn a platform-centric Coast Guard into the nucleus of a security web.

Here, the narrative accelerates. The focus shifts from what the Coast Guard hoped to build to what it has actually been forced to improvise, repurpose, and reinvent in the years since, as budget politics, organizational churn, and a rapidly deteriorating security environment reshaped every major element of its operational architecture.

This section examines two intimately connected stories.

The first is the tortured path from CN-295 to C-27J, a case study in how budget-driven politics, interservice trades, and episodic attention in Washington can upend seemingly settled aviation plans. The

Coast Guard's original maritime patrol aircraft roadmap envisioned a coherent, Deepwater-like family of platforms; what it received instead was a patchwork solution in which it had to absorb excess aircraft from another service, retrofit them for maritime use, and integrate them into an already strained personnel and logistics system. The result was not a carefully architected fleet, but a forced adaptation that nonetheless had to deliver real capability on day one, in an environment where gaps in maritime domain awareness translate quickly into lost lives, missed interdictions, and strategic opportunity for competitors and criminals alike.

The second story is the evolution of Coast Guard C4ISR from an aspirational Deepwater concept into an operational backbone that now underwrites virtually every mission set, from drug and migrant interdiction to Arctic operations, fisheries enforcement, and gray-zone competition in the Western Pacific.

In earlier chapters we saw how figures like Admiral Day and Admiral Currier understood, well before it was fashionable, that the service's real asymmetric advantage would come from networks, data fusion, and interoperability, not simply from new hulls or airframes.

Since then, that insight has been borne out: the Coast Guard's most effective cutters, aircraft, and shore facilities are those knitted into a broader web of national and international partners, able to turn fragmentary intelligence into targeted action, and to share a common operating picture from Washington to JIATF South to a boarding team in the eastern Pacific.

"Fast Forwarding" asks the reader to look at these developments not as discrete acquisition anecdotes, but as indicators of a deeper transformation in how the Coast Guard is being used and how it is forced to adapt.

The CN-295/C-27J story reveals a service perpetually adjusting to other people's budget deals, while still held to an "always ready" standard in the eyes of the American public. The C4ISR story shows a Coast Guard that has, in many respects, outpaced its better-resourced counterparts in embracing information-centric operations, even as it struggles to sustain and modernize the underlying infrastructure across an aging fleet and fragile shore plant.

Together, they illuminate the paradox at the heart of the modern Coast Guard: a service that is central to the nation's response to what I have elsewhere called the shift from crisis management to chaos management, yet one that remains structurally treated as an afterthought in budget and force-design debates.

By moving the narrative forward into these more recent and still-unfolding developments, Section VII connects the historical Coast Guard we have traced from the early Deepwater vision through 2016 to the Coast Guard now operating in a world of major power maritime rivalry, contested supply chains, and persistent gray-zone pressure.

It sets the stage for a frank assessment of what has and has not changed: the platforms are newer in some cases, the networks more capable, the mission set even broader, and the international expectations higher, but the underlying pattern of being persistently under-resourced has not been broken.

The chapters that follow use the CN-295 to C-27J transition and the C4ISR journey as lenses through which to evaluate whether the Coast Guard is being positioned to succeed in this new era or whether, once again, it is being asked to substitute ingenuity and dedication for coherent strategy and sustained investment.

# 1

# FROM CN-235 TO C-27J: HOW BUDGET POLITICS TRANSFORMED COAST GUARD AVIATION PROCUREMENT

IN THE COMPLEX world of defense acquisition, few stories better illustrate the unpredictable nature of military procurement than the U.S. Coast Guard's abrupt shift from the European-designed CN-235 maritime patrol aircraft to the Italian C-27J Spartan.

What began as a carefully planned modernization program under the Coast Guard's ambitious Deepwater initiative ended with the service inheriting fourteen C-27J aircraft that the Air Force no longer wanted, aircraft that were more expensive to operate but came at no acquisition cost.

This unexpected windfall fundamentally altered the Coast Guard's aviation fleet composition and serves as a compelling case study in how budget constraints, inter-service politics, and shifting strategic priorities can override years of systematic planning.

Within the larger Deepwater context, the Coast Guard's medium-range surveillance aircraft requirement represented a critical capability gap. The service needed to replace twenty-one aging Dassault HU-25 Falcon jets that had served faithfully but were becoming obsolete and expensive to maintain. The HU-25s, originally French business jets adapted for Coast Guard missions, lacked the endurance, cargo capacity, and modern sensor integration capabilities required for contemporary maritime patrol operations.

To meet this requirement, Integrated Coast Guard Systems awarded a contract to European Aeronautic Defense and Space Company (EADS) for a derivative of the CASA CN-235-300M aircraft. In Coast Guard service, this aircraft received the designation HC-144A Ocean Sentry. The selection represented the first major aviation asset procured under the Integrated Deepwater System program and embodied the program's philosophy of acquiring capable airframes equipped with state-of-the-art embedded command, control, surveillance, and reconnaissance systems.

Notably, the Lockheed team within ICGS explicitly rejected the C-27J in favor of the CN-235. They did so even though it was in the broader corporation's interest to recommend the C-27J. This showed the difference between a corporation acting as a lead systems integrator rather than as an organization selling its own products. It was rejected because the CN-235 fit the Deepwater systems approach much better than did the C-27J.

The CN-235 had a proven track record in military service worldwide. Originally developed jointly by Spain's CASA and Indonesia's IPTN as both a regional airliner and military transport, more than 230 CN-235s were operating globally by the time the Coast Guard made its selection.* The aircraft's twin-turboprop configuration, high-wing design, and rear loading ramp provided the versatility the Coast Guard required. Based on the CN-235-300 MP Persuader maritime patrol variant, the HC-144A offered significantly longer endurance than the HU-25 it replaced, up to eight to ten hours compared to the jet's shorter legs, along with superior performance in the critical low-level observation role.†

The HC-144A's design clearly addressed Coast Guard opera-

---

* https://www.militaryaerospace.com/home/article/16715987/eads-north-america-delivers-hc-144a-maintenance-training-unit-to-the-us-coast-guard
† https://www.military.com/equipment/hc-144a-ocean-sentry

tional requirements through several key features. Its large cargo cabin and rear ramp enabled mission tailoring on short notice. A standardized roll-on/roll-off palletized mission system facilitated rapid reconfiguration, allowing the aircraft to carry various sensor packages depending on mission requirements. The aircraft's relatively modest fuel consumption rate, combined with its short-field takeoff and landing capability, made it economically attractive for the service's dispersed operational requirements.

The Coast Guard's original procurement plan called for thirty-six HC-144A aircraft, with twelve Mission System Pallets that could be swapped between operational aircraft as missions required. This approach maximized flexibility while controlling costs, not all aircraft needed to carry the full sensor suite simultaneously.

Initial deliveries began in December 2006, with the program achieving Initial Operational Capability in April 2009. By January 2011, thirteen Ocean Sentry aircraft were operational with the Coast Guard, performing missions from air stations in Mobile, Alabama; Cape Cod, Massachusetts; and Miami, Florida.

The HC-144A quickly proved its worth in Coast Guard operations. The aircraft participated in high-profile missions including the Marquis Cooper search-and-rescue operation, response to the 2010 Haiti earthquake, environmental monitoring during the Deepwater Horizon oil spill, and Hurricane Sandy relief efforts. By June 2014, the Coast Guard's HC-144 fleet had accumulated 50,000 flight hours just five years after achieving IOC, a testament to the aircraft's reliability and mission effectiveness. The Ocean Sentry became the Coast Guard's most heavily utilized aircraft on a per-airframe basis, demonstrating the soundness of the original selection.[*]

Both the Coast Guard and industry observers anticipated steady

---

[*] https://www.defensemedianetwork.com/stories/multi-sensor-sentry/; https://defenceweb.co.za/sea/sea-sea/airbus-hc-144a-ocean-sentry-aircraft-fleet-surpasses-50-000-flight-hours-with-us-coast-guard/; https://www.aerocontact.com/en/aerospace-aviation-news/45811-airbus-hc-144a-ocean-sentry-aircraft-fleet-50-000-flight-hours-completed

production through the planned thirty-six aircraft buy. The fifteenth HC-144 delivered in June 2013, with the expectation that the program would continue through 2014 and beyond to complete the full procurement. The aircraft represented exactly what the Coast Guard needed: a reliable, economical, mission-flexible platform that could perform the service's diverse maritime patrol requirements without breaking the budget.

## The C-27J Spartan: The Air Force's Brief Affair

While the Coast Guard methodically built its HC-144 fleet, the U.S. Air Force pursued its own medium cargo aircraft program, one that would ultimately intersect with Coast Guard planning in unexpected ways. In 2007, the Pentagon awarded a $2.04 billion contract to Global Military Aircraft Systems (a joint venture of L-3 Communications and Alenia Aeronautica) for seventy-eight C-27J Spartan aircraft under the Joint Cargo Aircraft (JCA) program. The C-27J was designed to provide direct support to the U.S. Army, operated by Air National Guard units to deliver cargo the "last tactical mile" to fast-moving ground forces.

The C-27J Spartan traced its lineage to Italy's successful G.222 tactical transport. In 1995, Alenia and Lockheed Martin began discussions to modernize the G.222 design using components from the C-130J Hercules, including its advanced glass cockpit and more powerful engines. The resulting C-27J combined proven airframe design with cutting-edge avionics and propulsion, it shared engines and propellers with the larger C-130J, promising logistics commonality within the Air Force fleet.

During evaluation, both the C-27J and EADS CASA's C-295 (closely related to the Coast Guard's CN-235) competed for the JCA contract. The C-27J won based on its larger cargo capacity, wider and taller cabin dimensions, and compatibility with a broader range of military cargo. By November 2006, the aircraft had completed the

Department of Defense's Early User Survey evaluations, surpassing all requirements.

The first C-27Js entered Air National Guard service in 2010, with initial combat operations beginning in August 2011. Under Army tactical control in Afghanistan, the aircraft proved highly successful at the mission for which it was designed. The C-27J could access austere airfields that challenged larger C-130s while carrying substantially more cargo than helicopters, filling a genuine capability gap for deployed ground forces.[*]

However, this success story proved short-lived. The 2011 Budget Control Act and subsequent sequestration created intense pressure to reduce defense spending across all services. The Air Force, facing the need to cut $487 billion over ten years, began scrutinizing programs for potential savings. Despite the C-27J's operational effectiveness, the aircraft became a target for budget-driven cancellation.[†]

In early 2012, the Air Force announced plans not only to terminate further C-27J procurement but to retire the aircraft already delivered, a stunning reversal less than two years after the first aircraft entered service. Air Force Secretary Michael Donley justified the decision by arguing that the larger C-130 Hercules could fulfill ninety percent or better of Army needs without introducing a separate logistical supply chain and personnel training pipeline. The Air Force had conducted experiments in Iraq demonstrating that C-130s, though larger than optimal for some missions, could still perform direct support roles for ground forces.[‡]

The decision sparked fierce opposition from the Air National Guard, which had invested heavily in standing up C-27J units and saw the aircraft as ideally suited to Guard missions. Guard advocates disputed the Air Force's cost figures and questioned the logic of divesting new aircraft while retaining older, less capable platforms.

---

[*]  https://www.airandspaceforces.com/article/the-saga-of-the-spartans/

[†]  https://www.amarcexperience.com/ui/index.php?option=com_content&view=arti cle&id=53&catid=8&Itemid=159

[‡]  https://www.military.com/dodbuzz/2012/03/20/the-c-27-truth-vacuum

Various legislators also challenged the move, recognizing the waste inherent in acquiring and then immediately discarding modern aircraft. The controversy illustrated the tension between service-level budget priorities and the broader national interest in efficient defense spending.

Nevertheless, fiscal reality prevailed. The Air Force's analysis concluded that a C-27J cost approximately $308 million over its operational lifespan compared to $213 million for a 25-year C-130 lifecycle. While these figures were contested, the fundamental issue was that sustaining multiple aircraft types with separate logistics chains was unaffordable in the constrained budget environment. By 2013, newly built C-27Js were being delivered directly to the Davis-Monthan Air Force Base "boneyard" without ever entering operational service—a visible symbol of defense acquisition dysfunction.[*]

Ultimately, the Air Force spent $567 million acquiring twenty-one C-27Js between 2007 and 2013. Of these, seven would eventually transfer to U.S. Special Operations Command to replace aging CASA C-212 training aircraft. The remaining fourteen aircraft sat in preservation at Davis-Monthan, nearly new and operationally capable but declared excess to Air Force requirements. This created an unusual opportunity: high-quality military aircraft available for inter-service transfer at zero acquisition cost.

## The Unexpected Transfer: Budget Windfall or Trojan Horse?

When the Air Force made clear its intention to divest the C-27J fleet, Coast Guard leadership immediately recognized both an opportunity and a dilemma. In July 2013, the Coast Guard began seriously considering acquiring up to fourteen retired C-27Js and converting

---

[*] https://www.military.com/dodbuzz/2012/03/19/the-air-force-numbers-game; https://www.military.com/dodbuzz/2012/03/13/ohio-guard-accuses-af-of-fudging-c-27j-figures; https://www.af.mil/News/Article-Display/Article/111442/c-27-program-cut-explained-budget-aligned-with-strategy/

them for search-and-rescue missions while canceling undelivered orders for the HC-144 Ocean Sentry. Budget projections suggested this approach could save between $500 million and $800 million compared to completing the full thirty-six aircraft HC-144 procurement.[*]

Coast Guard Commandant Admiral Robert Papp publicly endorsed the transfer, stating that acquiring the fourteen C-27Js at no cost would save the Coast Guard approximately half a billion dollars in lifecycle costs. The political and budgetary logic was compelling: why spend money procuring new HC-144s when serviceable C-27Js were available for free? In an era of constrained budgets and sequestration, the transfer seemed like common-sense efficiency.[†]

However, the proposal immediately sparked controversy. EADS, manufacturer of the HC-144, vigorously contested the Coast Guard's analysis. The company argued that the HC-144 was half as expensive to maintain and operate compared to the C-27J in terms of direct maintenance and fuel costs, calling into question whether the transfer would actually generate the projected savings.

This was one of the key reasons Lockheed as an LSI had rejected the C-27J in favor of the HC-144.

The C-27J was a larger, more complex aircraft with higher performance but correspondingly higher operating costs. Its commonality with C-130J engines provided some logistics advantages, but the Coast Guard didn't operate C-130Js: it operated HC-130Hs and HC-130Js, and the engine commonality argument was less relevant than it would have been for the Air Force.

The debate revealed fundamental tensions in defense acquisition strategy. From a pure acquisition cost perspective, the C-27J transfer was undeniably attractive for the aircraft came free, eliminating the most visible portion of procurement expense.

---

[*]   https://cgaviationhistory.org/2014-coast-guard-acquires-c-27j-aircraft/

[†]   https://theaviationist.com/2013/11/19/uscg-wants-c-27j/

But military aviation professionals understand that acquisition cost typically represents only twenty to thirty percent of an aircraft's total lifecycle expense. Operating costs, fuel, maintenance, spare parts, specialized training, infrastructure modifications constitute the majority of long-term expense. The question was whether "free" aircraft that cost more to operate would actually prove more economical than continuing with the purpose-designed, lower-operating-cost HC-144 program.

Coast Guard leaders also emphasized other advantages of the C-27J. The aircraft offered greater range, speed, and payload capacity than the HC-144. Its cargo capacity could prove valuable for disaster response missions, where the Coast Guard frequently delivers supplies to affected areas. The C-27J's higher cruise speed could reduce transit time to distant search-and-rescue cases. And as Admiral Papp noted, the C-27J was "one of the aircraft we looked at when we started the Deepwater project," suggesting it had always been considered a viable alternative to the CN-235/HC-144.[*]

Congress ultimately resolved the debate by including the transfer in the Defense Authorization Bill for Fiscal Year 2014, signed on December 26, 2013. The Coast Guard received authorization to accept fourteen C-27Js from the Air Force, fundamentally altering the service's medium-range aviation strategy. In response, the Coast Guard halted HC-144 procurement, with the eighteenth and final Ocean Sentry delivered on October 7, 2014, exactly half the originally planned thirty-six aircraft fleet.

## Implementation Challenges: The Hidden Costs of "Free" Aircraft

While the C-27J transfer offered undeniable acquisition savings, the Coast Guard quickly discovered that operationalizing the aircraft

---

[*]   https://www.dcms.uscg.mil/Portals/10/CG-9/Acquisition%20PDFs/Factsheets/C27J.pdf

would prove far more complex and expensive than initially anticipated. Although the aircraft came at no procurement cost, a January 2015 Government Accountability Office report revealed that the Coast Guard estimated needing approximately $600 million to fully operationalize the fourteen C-27Js, with plans to field fully operational aircraft extending to 2022.[*]

The complexity of the integration process created multiple challenges.

- First, the C-27Js delivered to the Coast Guard came configured as cargo aircraft, not maritime patrol platforms. Unlike the purpose-built HC-144 Ocean Sentry with its integrated sensor suite and mission systems, the C-27Js required extensive modification, a process called "missionization" which was necessary to perform Coast Guard missions. This involved adding electro-optical/infrared sensors, multimode radar, bubble observation windows, communications equipment, and the Minotaur mission system architecture that the Coast Guard was implementing across its fixed-wing fleet.
- Second, the Coast Guard lacked immediate access to critical technical data required for modifications. The aircraft's structural modifications, particularly those needed to incorporate radar and sensor systems, required manufacturer technical documentation that wasn't automatically transferred with the aircraft themselves.[†] This created potential delays and complications in the modification process, as the Coast Guard couldn't proceed with structural changes without proper engineering data.
- Third, establishing a spare parts inventory proved more

---

[*]  https://www.gao.gov/products/gao-15-325
[†]  https://www.gao.gov/products/gao-15-325

challenging than expected. The Coast Guard needed to purchase a complete set of spare parts for each aircraft to ensure operational availability. However, the service faced potential pricing issues and delivery delays from the manufacturer, as the C-27J production line was winding down following the Air Force cancellation. Limited continuing production complicated parts procurement and raised concerns about long-term sustainability.

- Fourth, the integration required establishing entirely new maintenance infrastructure and training programs. Unlike the HC-144, which shared some logistics commonality with the Coast Guard's existing EADS/Airbus helicopter fleet, the C-27J was an entirely new platform requiring separate maintenance training, technical publications, ground support equipment, and specialized tools. Air Station Sacramento, California, was designated as the first operational C-27J unit, requiring substantial facility modifications and equipment installation before aircraft could deploy.

The HC-27J missionization timeline stretched considerably beyond early projections. The Coast Guard, working with Naval Air Systems Command, sent the first C-27J "Spartan" to Naval Air Station Patuxent River, Maryland, where it entered the mission system integration process in September 2017. Coast Guard acquisition documents report that prototype installation was not completed until 2022, with testing and verification still ongoing after that point.

This meant that nearly a decade elapsed between Congress directing the transfer of 14 C-27Js to the Coast Guard in the Fiscal Year 2014 National Defense Authorization Act and the emergence of a missionized prototype approaching operational capability. That drawn-out timeline contrasted sharply with the "no-cost" and "faster and cheaper" logic used by Coast Guard leadership to justify capping the HC-144 Ocean Sentry fleet at 18 aircraft in favor of regenerating and missionizing the transferred Spartans.[*]

Despite these challenges, the C-27J did offer capability advantages over the HC-144. The aircraft's performance characteristics, including up to twelve hours of endurance in the operational theater, higher speed, and greater payload capacity, made it well-suited for the Coast Guard's evolving mission set. The service began operating C-27Js from Air Station Sacramento and later from Air Station Elizabeth City, North Carolina. By 2016, the aircraft had already logged over 1,750 flight hours, including operations in support of law enforcement, search-and-rescue missions, and various other activities.[†]

By May 1, 2017, Air Station Sacramento received its sixth and final C-27J, giving the unit its full complement of aircraft. The Coast Guard designated the modified aircraft as HC-27J, distinguishing the Coast Guard maritime patrol variant from the Air Force's cargo configuration. The HC-27J incorporated the full Minotaur mission system with surface search radar, electro-optical/infrared sensors, comprehensive C4ISR capabilities, and observer windows designed to enable aircraft crews to detect, classify, and identify maritime targets.[‡]

## Financial and Strategic Implications

The shift from HC-144 to C-27J procurement created significant ripple effects throughout Coast Guard aviation planning. A 2015 GAO analysis examined how the change affected both costs and

---

[*]  https://wildfiretoday.com/gao-evaluates-the-coast-guards-acquisition-of-c-27j-aircraft/

[†]  https://www.dcms.uscg.mil/Our-Organization/Assistant-Commandant-for-Acquisitions-CG-9/Programs/Air-Programs/MRS-HC-144-C27J/C-27J-Program-Profile/; https://www.leonardo.com/en/news-and-stories-detail/-/detail/the-u.s.-coast-guard-s-c-27j

[‡]  https://www.dcms.uscg.mil/Our-Organization/Assistant-Commandant-for-Acquisitions-CG-9/Programs/Air-Programs/Minotaur-Mission-System/

operational capability. The findings were nuanced and somewhat sobering.

The GAO confirmed that the C-27J would indeed improve the affordability of the Coast Guard's fixed-wing fleet compared to the original Deepwater program of record, with estimated savings of $837 million over thirty years.

However, the analysis revealed that the source of these savings had shifted fundamentally from what Coast Guard leadership initially projected. Rather than savings from lower acquisition costs being applied to maintain the same flight hour capacity, a significant portion of savings actually resulted from an eighteen percent reduction in total flight hours due to changes in the planned aircraft mix.[*]

This finding highlighted a critical issue: the Coast Guard was achieving cost savings partly by reducing operational capacity, not solely through more efficient procurement. The service's baseline requirement called for 52,400 flight hours annually to accomplish its missions.[†] The shift to a smaller combined fleet of HC-144s and HC-27Js, even with the C-27J's greater individual capability, meant fewer total flight hours available than originally planned. Whether this reduced capacity adequately covered the Coast Guard's expanding mission requirements remained an open question.

The GAO recommended that the Coast Guard conduct a comprehensive fleet mix analysis before accepting additional HC-130J aircraft, as premature expansion of the heavy aircraft fleet might result in excess capacity or suboptimal mission coverage. The recommendation reflected concern that the service was making tactical procurement decisions without fully analyzing strategic fleet requirements.

The Department of Homeland Security agreed with the recommendation for a fleet study but disagreed with delaying HC-130J

---

[*]   https://www.gao.gov/products/gao-15-325
[†]   https://www.airandspaceforces.com/article/the-saga-of-the-spartans/

procurement, arguing that the Coast Guard's understanding of its requirements was sufficient to proceed.

From a strategic perspective, the C-27J transfer illustrated both the opportunities and pitfalls of inter-service equipment transfers.

On one hand, the transfer prevented waste: rather than scrapping nearly-new aircraft, the C-27Js found productive employment in Coast Guard service. This represented good stewardship of taxpayer resources and demonstrated the kind of inter-service cooperation that defense reform advocates frequently encourage.

On the other hand, the transfer highlighted how short-term budget pressures can override systematic capability planning. The Coast Guard had conducted a deliberate analysis under the Deepwater program, evaluating multiple aircraft options before selecting the HC-144. That aircraft was performing well, accumulating flight hours faster than any other Coast Guard platform, and proving its operational effectiveness across diverse missions.

Yet budget considerations forced abandonment of the carefully planned program in favor of a more complex, higher-operating-cost solution that happened to be available at no acquisition cost.

The episode also demonstrated the difficulties of accurate life-cycle cost estimation in defense acquisition. Both the Air Force's initial C-27J cost projections and the Coast Guard's estimates of operationalization expenses proved optimistic. The Air Force's claim that C-27J lifetime costs were $308 million versus $213 million for C-130s was disputed by the Air National Guard.

Similarly, the Coast Guard's projection of $600 million to operationalize fourteen C-27Js understated the complexity and duration of the integration effort. These estimation challenges reflect inherent uncertainties in defense acquisition but also suggest that budget-driven decisions sometimes rely on overly optimistic assumptions.

## Lessons and Broader Implications

The Coast Guard's unexpected pivot from CN-235/HC-144 to C-27J procurement offers several important lessons for defense acquisition policy and practice.

- First, the episode illustrates the profound impact of budget instability on military planning. The Air Force's rapid reversal on the C-27J, from urgent requirement to immediate divestment in less than three years, reflected the chaotic budget environment created by sequestration and continuing resolutions. When services lack predictable, multi-year funding, acquisition programs become vulnerable to sudden cancellation regardless of operational merit.
- Second, the case demonstrates how "free" is not always cheap. The C-27J's zero acquisition cost made the transfer politically irresistible, but the extensive modifications, training infrastructure, spare parts procurement, and ongoing operating costs created substantial expense that partially offset the acquisition savings. Defense decision-makers must resist the temptation to focus exclusively on acquisition costs when evaluating alternatives, as lifecycle operating and support costs typically dominate total ownership expense.
- Third, the experience shows the value of purpose-designed solutions over adapted alternatives. The HC-144 was specifically configured for Coast Guard maritime patrol missions from the outset, with mission systems integrated during production and operating characteristics optimized for the service's requirements. The C-27J, though a capable aircraft, required years of expensive modification to perform missions for which it wasn't originally designed. Purpose-built solutions, while

potentially more expensive initially, often prove more cost-effective over time than adapted alternatives.

- Fourth, the case highlights the importance of maintaining adequate fleet capacity, not just individual platform capability. While the C-27J offered superior performance compared to the HC-144 in speed, range, and payload, the smaller combined fleet size reduced total available flight hours. Military capability derives not only from platform performance but also from sufficient capacity to maintain persistent coverage across required areas of operation. Trading fleet size for individual platform capability involves real operational risk.
- Fifth, the episode demonstrates both the benefits and risks of inter-service transfers. The C-27J transfer prevented outright waste of taxpayer investment and provided the Coast Guard with capable aircraft. However, the transfer also disrupted the Coast Guard's systematic aviation modernization, creating integration challenges and forcing abandonment of a working program. Inter-service transfers work best when receiving services have sufficient time and resources to properly integrate transferred equipment rather than feeling compelled to accept it purely for budget reasons.
- Finally, the experience underscores the need for realistic cost estimation and schedule projection in defense acquisition. Both the Air Force's initial C-27J program and the Coast Guard's integration effort encountered cost growth and schedule delays beyond original projections. More conservative, realistic planning, acknowledging the typical challenges of defense acquisition, would better prepare services and policymakers for the true resource requirements of acquisition programs.

In short, the Coast Guard's shift from CN-235/HC-144 Ocean

Sentry procurement to C-27J Spartan adoption represents a fascinating case study in how defense acquisition plans can be fundamentally altered by factors external to the original requirement. What began as a methodical, requirements-driven selection of a purpose-designed maritime patrol aircraft evolved into opportunistic acceptance of excess aircraft from another service's cancelled program.

The HC-144 Ocean Sentry proved itself an effective, economical platform well-suited to Coast Guard missions. The eighteen aircraft delivered continue to serve reliably, accumulating flight hours at rates that exceed other Coast Guard aircraft types. Yet budget constraints forced premature termination of the program at half the planned fleet size.

The C-27J Spartan, despite never being originally intended for Coast Guard missions, brought genuine capabilities to the service. After extensive modification and years of integration effort, the HC-27J has become an operational asset contributing to search-and-rescue, law enforcement, and disaster response missions. The aircraft's superior speed, range, and payload capacity offer advantages over the HC-144 for certain mission profiles.

Whether the Coast Guard achieved genuine savings through this substitution remains debatable. The free acquisition cost saved hundreds of millions of dollars, but the extensive missionization requirements, reduced fleet flight hours, and higher operating costs partially offset those savings. The ten-year timeline from transfer authorization to full operational capability contrasted sharply with the "faster and cheaper" promise that justified the change.

Perhaps most significantly, the episode illustrates how budget-driven decision-making can override careful planning and disrupt systematic modernization programs. The Coast Guard's Deepwater initiative, despite its flaws and challenges, represented an attempt at comprehensive, long-term fleet planning. The C-27J transfer, while financially opportunistic, injected unpredictability into that planning process and forced acceptance of a solution that hadn't been chosen through systematic requirements analysis.

In the end, the Coast Guard operates both HC-144s and HC-27Js as part of its medium-range surveillance fleet, with each aircraft type bringing distinct capabilities and operating characteristics. The service adapted successfully to the unexpected change in acquisition strategy, demonstrating the flexibility and resourcefulness for which the Coast Guard is known.

Yet the story serves as a cautionary tale about the challenges of defense acquisition in an era of budget instability, the hidden costs of "free" equipment, and the difficulty of maintaining coherent modernization programs when short-term financial pressures override long-term strategic planning.

The Coast Guard's pivot from CN-235 to C-27J ultimately says less about either aircraft's merits than about the broader dysfunction in U.S. defense acquisition processes. When careful planning can be overturned by inter-service budget shuffling, when "free" aircraft require half a billion dollars to operationalize, and when services must accept equipment they didn't choose simply because budget constraints make any alternative unaffordable, the system is revealing fundamental weaknesses that go well beyond individual program decisions.

The Coast Guard's HC144 Ocean Sentry, shares the ramp at the Aviation Logistics Center in Elizabeth City with an HU25 Falcon and the long-retired HU16 Albatross. USCG Photo/Dave Silva.

A Coast Guard Air Station Clearwater C27J Spartan and MH60 Jayhawk flyover the Blackthorn memorial in St. Petersburg, Florida on Jan. 28, 2025. 45 years ago, Blackthorn collided with the motor vessel Capricorn near the Skyway Bridge and 23 members of the crew were lost. (U.S. Coast Guard photo by Petty Officer 1st Class Riley Perkofski)

A Coast Guard HC144 Ocean Sentry, In flight
over the Albemarle Sound in North Carolina.
USCG Photo/ Dave Silva, 2009

A U.S. Coast Guard C-130 Hercules with USCG
Base Honolulu, performs aerial maneuvers while
performing a Search and Rescue demonstration
during the 2025 Kaneohe Bay Air Show at
Marine Corps Base Hawaii, Aug. 10, 2025. U.S.
Marine Corps photo by Cpl. Jade K. Venegas.

# 2

# FROM CRISIS TO TRANSFORMATION: THE COAST GUARD'S C4ISR JOURNEY SINCE ADMIRAL DAY'S WARNING

WHEN REAR ADMIRAL Robert E. Day Jr. sat down for an interview in 2010 about the Coast Guard's C4ISR capabilities, he painted a picture of a service wrestling with fundamental challenges in command, control, communications, computers, intelligence, surveillance, and reconnaissance.

Admiral Day had just assumed duties as the Assistant Commandant for Command, Control, Communications, & Information Technology and Director of the Coast Guard Cyber Command Pre-Commissioning Detachment, stepping into a role that would prove prophetic in highlighting the technological challenges ahead.

His candid assessment of the Coast Guard's information technology needs particularly his discussion of systems like Watchkeeper barely surviving on life support and the service's struggle to fund essential C4ISR capabilities foreshadowed a crisis that would eventually force the service to embark on one of the most ambitious technology modernization efforts in its history.

Admiral Day's interview captured a moment when the Coast Guard was beginning to understand that C4ISR was not merely a supporting capability but the essential nervous system enabling modern maritime operations. He emphasized that without effective ISR systems, the effectiveness of personnel and resources would be fundamentally undercut, and that efficient C4ISR systems were essential to mission success. His example of drug interdiction operations in the Eastern Pacific illustrated the stark contrast between the old days of "boring holes in the water" and hoping for lucky encounters versus modern operations enabled by intelligence-driven deployments and sophisticated surveillance systems.

Yet even as Admiral Day articulated this vision, funding remained the critical constraint. He noted that many needed changes over the next five years "may or may not occur, because they may or may not make the funding threshold. In most cases right now, they are not going to make the funding threshold."

This frank acknowledgment of resource limitations proved prescient, as the Coast Guard would spend the subsequent decade and a half struggling with chronic underinvestment in information technology that would bring the service to what one commandant would later call "the brink of catastrophic failure."

## The Years of Underinvestment: 2010-2019

The decade following Admiral Day's interview saw the Coast Guard making incremental progress on some C4ISR initiatives while falling further behind on others. The service continued deploying key systems that Admiral Day had highlighted, including the Rescue 21 coastal distress and safety system, the Nationwide Automatic Identification System (NAIS) for vessel tracking, and the Watchkeeper common operational picture system. By 2013, testimony before Congress revealed that Watchkeeper had been provided to 26 of 35 locations and had proven operationally effective, benefiting from Coast Guard and other agency blue force tracking capabilities.[*]

---

[*] https://www.congress.gov/113/chrg/CHRG-113hhrg82266/CHRG-113hhrg82266.pdf

During this period, the Coast Guard focused on developing a Common Operational Picture as the central component of effective C4ISR. This system was designed to receive inputs from disparate information sources, process and correlate the data, and distribute it through the enterprise across multiple security domains. The Command and Control Personal Computer (C2PC) became the most prevalent C2 system in the C4ISR architecture, allowing viewing of COP data on standard Coast Guard workstations. The Enterprise Geospatial Information System (E-GIS) created a storehouse of information, including maps and charts originally developed to support the viewing and exchange of geospatial data.

The Coast Guard's C4ISR modernization project during this era was structured as a multi-year, multi-billion dollar effort to design, develop, and integrate electronic equipment and computer systems aboard the service's newest assets, including National Security Cutters, Offshore Patrol Cutters, HC-130J long-range surveillance aircraft, and HC-144A and C-27J medium-range surveillance aircraft. The program acquired and integrated electronic sensors, networking, and data processing capabilities across the fleet.

However, a 2014 Government Accountability Office report revealed troubling realities: the Coast Guard had changed its testing strategy for C4ISR systems and was no longer planning to test key requirements. The service had repeatedly delayed and reduced capability through its annual budget process, and without a long-term fleet modernization plan, it could not determine the extent to which it would meet mission needs and achieve desired results.*

A 2015 Department of Homeland Security Inspector General report found that while delivered C4ISR capabilities supported mission needs on some ships and aircraft, the Coast Guard had not carried out planned system enhancements necessary to support the mission needs of certain aircraft and legacy ships. Budget reductions

---

* https://transportation.house.gov/uploadedfiles/2014-06-18-mackin.pdf, https://www.files.ethz.ch/isn/181272/663784.pdf

had forced the service to scale back planned upgrades to aging technology on legacy cutters and cancel planned major upgrades to aircraft C4ISR mission systems. As a result, Coast Guard personnel aboard these platforms continued to rely on obsolete technology that negatively affected mission performance and made operations and maintenance more difficult and costly.[*]

## The Crisis Point: The Technology Revolution Roadmap

By early 2020, the accumulated effects of years of deferred investment had brought the Coast Guard's information technology infrastructure to a critical juncture. In a State of the Coast Guard address delivered in South Carolina in February 2020, Commandant Admiral Karl Shultz announced what he called a "tech revolution," acknowledging bluntly that "years of investment tradeoffs have brought our information technology to the brink of catastrophic failure."[†]

This stark assessment, far more dire than anything Admiral Day had articulated a decade earlier, reflected the reality that the service's email servers routinely failed, cutters deployed without internet connectivity, and teams across the organization could not easily collaborate.

Admiral Shultz unveiled an ambitious Technology Revolution Roadmap with goals that seemed almost fantastical given the current state of the service's IT infrastructure: doubling bandwidth on cutters off the U.S. shoreline, boosting internet speeds by a factor of fifty, and fundamentally modernizing the Coast Guard's approach to technol-

---

[*]  https://www.nationaldefensemagazine.org/articles/2014/12/1/2014december-coast-guard-information-technology-sensor-needs-go-unfilled; https://www.nationaldefensemagazine.org/articles/2014/12/1/2014december-coast-guard-information-technology-sensor-needs-go-unfilled

[†]  https://www.nationaldefensemagazine.org/articles/2020/4/24/coast-guard-pursuing-ambitious-tech-revolution; https://fedscoop.com/coast-guard-tech-revolution-plan/

ogy. The roadmap emphasized that "the Coast Guard of tomorrow must operate beyond brick and mortar" and called for closing a $300 million IT spending shortfall in the budget.

The Technology Revolution focused on five primary lines of effort: modernizing C5I (the Coast Guard had added "Cyber" to the traditional C4ISR formulation) infrastructure, improving cutter connectivity, improving cyber readiness, transitioning to modern phone systems, and transitioning to modern software that provides mobility and leverages cloud technology. The initiative also emphasized moving to an "industry standard" replacement cycle for technology to avoid repeating the same pattern of deferred maintenance that had created the current crisis.

The FY 2022 budget request reflected these priorities, with just over $99 million allocated to support the Technology Revolution Roadmap. Acquisitions funding included $18 million for C4ISR systems and a $2 million boost for cyber and enterprise mission platforms. The budget emphasized IT and cybersecurity investments to support five primary lines of effort, with particular focus on cutter connectivity improvements.

## Transformation in Action: 2020-2024

The years following the Technology Revolution announcement saw significant progress, though implementation faced ongoing challenges. By 2024, Coast Guard personnel could point to tangible improvements: cutters that once were limited to low-bandwidth message traffic now had connectivity sufficient for video conferencing while underway. The service made substantial strides in improving underway connectivity, with investments beginning to pay dividends across the fleet.

The organizational structure supporting C4ISR evolved significantly during this period. The establishment of the C5I Service Center (C5ISC), expanding the traditional C4ISR acronym to explicitly include "Cyber", created a dedicated organization with a mission of delivering technical solutions for mission success. The C5ISC oversees more than 350 systems across the Coast Guard enterprise, providing essential situational awareness, data processing, and information exchange tools that enhance all aspects of mission operations. The center also provides subject matter experts to advise Coast Guard organizations on emerging technologies like artificial intelligence and the implementation of zero-trust cybersecurity architecture.[*]

The service's approach to C5I modernization became increasingly data-centric, recognizing that artificial intelligence, machine learning, and other emerging technologies all depend on effective data management, data security, and data mobility. The C5ISC began implementing zero-trust architecture across its systems, working to create an integrated data environment that would eliminate siloed backend databases and enable big data processing.[†]

This represented a fundamental shift from the fragmented systems Admiral Day had described in 2010, moving toward an enterprise approach that would enable the sophisticated analytics and decision support that modern operations demand.

Contract actions during this period reflected the scale and complexity of C5I modernization efforts.

In June 2024, the Coast Guard selected eight small businesses for the second iteration of a potential $233.3 million multiple-award contract to provide installation and logistics management support services for C5I systems. The Installation and Logistics Management Services II contract focused on the implementation, sustainment, and disposition of C5I systems across both shipboard and land-based

---

[*]  https://www.choisystechnology.com/contract-vehicles/uscg-cediss/; https://www.dvidshub.net/video/888638/coast-guard-c5isc-information-assurance-branch-customer-industry-overview-presentation

[†]  https://www.dcms.uscg.mil/Portals/10/CG-9/Acquisition%20PDFs/RDTE/Current%20RDC%20Research%20Program%20Portfolio.pdf?ver=hMgKj7e_haksSdLPUgBaPw%3D%3D; https://www.nationaldefensemagazine.org/articles/2023/12/26/coast-guard-goes-all-in-on-zero-trust-architecture

installations. In April 2025, Chugach Government Solutions announced it had secured a $164 million contract from Naval Information Warfare Center Atlantic to provide C5I system integration, engineering, and technical support for U.S. Coast Guard C5I modernization, continuing more than a decade of support.[*]

However, challenges persisted. A 2024 analysis in the U.S. Naval Institute's Proceedings assessed that while the Technology Revolution had achieved significant advances in areas like cutter connectivity, it had fallen short in other critical areas. The authors argued that the Coast Guard needed to integrate new lines of effort to prepare for the future, including establishing a Software Factory similar to those created by other military services, expanding the Coast Guard Academy's graduate education programs to include data science and AI, formalizing its approach to unmanned systems development and deployment, and institutionalizing process automation capabilities.[†]

## The Robotics and Autonomous Systems Revolution

Perhaps the most dramatic departure from the C4ISR paradigm Admiral Day described came with the establishment of the Robotics and Autonomous Systems Program Executive Office in August 2025. This organizational innovation, described as potentially the most transformational enhancement to capability since the inception of aviation, reflects the Coast Guard's recognition that unmanned and autonomous systems represent not merely incremental improvements but a fundamental shift in how maritime operations will be conducted.

---

[*]  govconwire.com/2024/06/coast-guard-awards-8-spots-on-233m-follow-on-c5i-system-installation-logistics-management-support-idiq/;   https://www.chugachgov.com/2025/04/07/coast-guard-c5i-modernization-contract-niwc-atlantic/

[†]  https://www.usni.org/magazines/proceedings/2024/august/where-coast-guards-techrev-fell-short-and-path-new-one

The RAS PEO is dedicated to the rapid operationalization of the Coast Guard's Unmanned Systems Strategic Plan, with authority to look across surface, subsurface, counter-UAS, and shore/medium/long-range UAS capabilities. In September 2025, the Coast Guard announced a nearly $350 million investment in robotics and autonomous systems under the One Big Beautiful Bill Act, with $11 million in fiscal year 2025 for immediate system upgrades. These initial investments included $4.8 million to procure 16 VideoRay Defender remotely operated vehicles to replace Deployable Specialized Forces' aging ROV fleet, $2 million to procure Qinetiq Squad Packable Utility Robots and mini-SPURs to replace outdated unmanned ground vehicles at Strike Teams, and $4.3 million to purchase 125 SkyDio X10D short-range unmanned aircraft systems.[*]

This emphasis on robotics and autonomous systems represents a recognition that future C5I architectures must integrate not only human-operated platforms but increasingly sophisticated autonomous and semi-autonomous systems operating across multiple domains. The data management, processing, and decision support requirements for coordinating such systems represent a quantum leap beyond what Admiral Day described in 2010, requiring new approaches to AI-enabled mission systems, edge computing, and distributed decision-making.

## Cyber as a Core Mission

The evolution from C4ISR to C5ISR with the explicit addition of "cyber" reflects more than a change in nomenclature. It represents the Coast Guard's recognition that cybersecurity is no longer a subset of information technology management but a core operational impera-

---

[*] https://www.news.uscg.mil/Press-Releases/Article/4314137/coast-guard-to-invest-350-million-in-robotics-and-autonomous-systems/; https://www.executivegov.com/articles/uscg-peo-robotics-autonomous-systems-launch

tive. The service established Coast Guard Cyber Command and has worked to implement comprehensive zero-trust architecture across its enterprise systems. The FY 2022 budget included funding to establish a third Cyber Protection Team to work with cyber specialists at critical ports of entry, recognizing the vulnerability of maritime critical infrastructure to cyber threats.[*]

The emphasis on zero-trust architecture and data-centric approaches reflects lessons learned from the fragmented systems that plagued earlier C4ISR efforts. Rather than defending perimeter networks, zero-trust assumes no implicit trust and requires continuous verification of users and systems. This approach is particularly critical for an organization like the Coast Guard that operates across civilian and military networks and must share data with numerous federal, state, and local partners.

## Modern Operational Picture: Beyond Admiral Day's Vision

When Admiral Day discussed the Common Operational Picture in 2010, he emphasized systems like Watchkeeper providing sector-level decision support and situational awareness. Today's operational picture extends far beyond those early systems. The Coast Guard's modern C5ISR architecture must integrate data from an expanding array of sensors, including space-based ISR systems, sophisticated radar and electronic warfare systems aboard National Security Cutters and Offshore Patrol Cutters, data from autonomous systems operating independently or in swarms, artificial intelligence-enabled analytics identifying patterns in vast datasets, and networked communications enabling real-time collaboration across the service and with partners.

The Offshore Patrol Cutter program, with its Northrop Grumman-provided C5ISR systems, exemplifies the modern approach.

---

[*] https://www.csis.org/analysis/us-coast-guard-and-future-maritime-cybersecurity

These systems include an advanced computing "nerve center," sophisticated data-sharing capabilities, modern communication systems, and integrated bridge systems. The goal is not merely to provide better versions of existing capabilities but to enable fundamentally new operational concepts, including distributed maritime operations in contested environments, seamless interoperability with Navy and Marine Corps forces, AI-assisted decision support for commanding officers, and real-time integration of data from autonomous systems operating in support of manned platforms.[*]

## Challenges and the Road Ahead

Despite significant progress, the Coast Guard faces ongoing challenges in C5ISR modernization. The October 2025 U.S. Naval Institute Proceedings article on fleet modernization noted that 52 percent of scheduled dry-dock periods were canceled in fiscal year 2024, pushing back more than $219 million worth of maintenance. This affects not only hull, mechanical, and electrical systems but also the installation and maintenance of C5ISR equipment. The service continues to balance immediate operational demands against long-term modernization requirements, a tension Admiral Day identified in 2010 that has not disappeared despite increased resources.[†]

The rapid pace of technological change presents its own challenges. Technologies that seem cutting-edge today may be obsolete within the decade-long timelines typical of major acquisition programs. The Coast Guard's approach to this challenge, exemplified by the Technology Revolution's emphasis on industry-standard replacement cycles and Force Design 2028's focus on agility and adaptability (discussed in the next chapter). represents a learning

---

[*] https://maritime-executive.com/corporate/northrop-grumman-to-develop-c5isr-control-systems-for-us-coast-guard;  https://warriormaven.com/news/sea/video-special-inside-the-new-u-s-coast-guard-offshore-patrol-cutter

[†] https://www.usni.org/magazines/proceedings/2025/october/how-modernize-coast-guard-fleet

process that extends back to Admiral Day's era but has matured significantly.

Workforce development remains critical. The Coast Guard must attract, develop, and retain personnel with expertise in cybersecurity, data science, artificial intelligence, software development, and the integration of autonomous systems. The service's partnership with the Defense Innovation Unit and exploration of concepts like Software Factories reflect recognition that technology modernization is as much about people and processes as it is about hardware and software.

Looking back at Admiral Day's 2010 interview from the vantage point of late 2025, several themes emerge.

- First, his warning about funding thresholds proved prescient. The Coast Guard did indeed spend years unable to adequately resource its C4ISR needs, leading to the crisis Commandant Shultz described in 2020.
- Second, his emphasis on C4ISR as essential rather than merely supporting proved exactly right. The organizational elevation of C5ISR to a dedicated PEO under Force Design 2028 institutionalizes this reality.
- Third, his vision of decision support tools and Common Operational Picture capabilities, while forward-thinking in 2010, has been far exceeded by the AI-enabled, data-centric, autonomous-system-integrated C5I architecture the Coast Guard is now building.

The journey from Admiral Day's modest hopes for systems like Watchkeeper to the Technology Revolution and Force Design 2028 has been neither smooth nor assured. It required a crisis to galvanize resources and attention.

But with more than $24.5 billion in modernization funding, dedicated organizational structures, and a clearer strategic vision than at any point in the past fifteen years, the Coast Guard appears positioned to realize the C4ISR capabilities Admiral Day articulated as essential to a 21st century force.

The challenge now is sustaining this progress beyond the current leadership and funding cycle, ensuring that the next generation of Coast Guard leaders does not find themselves where Admiral Day did in 2010, trying to sustain critical systems "on life support" while arguing that investments "may or may not make the funding threshold."

# PART 8

---

# CONCLUSION

Part eight returns the reader to the central question that has run, often implicitly, through this entire book: what kind of Coast Guard does the United States actually need for the world it now inhabits, and what kind is it willing to pay for and integrate into its broader defense and security architecture?

Up to this point, the chapters have largely been empirical and episodic: commandant perspectives, district-level realities, the rise and fall of Deepwater, the hard-won success of the National Security Cutter, the evolution of maritime patrol aviation and C4ISR, and the services pivotal role in crisis response from Katrina to Deepwater Horizon.

Section VIII steps back from the narrative flow and treats that accumulated experience as strategic evidence. It asks what these episodes, taken together, reveal about the Coast Guard's enduring strengths, its structural vulnerabilities, and the choices facing policy makers as the United States moves deeper into an era of persistent competition and what I have called chaos management.

The first chapter, "The Coast Guard at a Strategic Crossroads: A

2011 Perspective," deliberately revisits an earlier vantage point to show how clearly many of today's dilemmas were already in view a decade and a half ago. At that moment, Deepwater's conceptual promise had collided with fiscal and bureaucratic reality, the service was being pulled between homeland security and traditional missions, and its away-game responsibilities from the Arctic to the Western Pacific were expanding without commensurate resources.

What is striking, reading that period now, is not how much has changed, but how much has not. The same themes recur: aging platforms, fragile infrastructure, chronic personnel shortfalls, and a mission set that grows by accretion rather than by conscious design. By returning to 2011, this chapter underscores that today's crossroads is not a sudden development but the product of long-running trends that successive administrations have chosen to manage, rather than resolve.

" The Hope of Force Design 2028" shifts the lens to a more recent conceptual effort to rethink how U.S. forces including, at least on paper, the Coast Guard might be organized, equipped, and employed in a world of distributed operations, integrated deterrence, and gray-zone competition. For the Navy and Marine Corps, force design conversations have focused on new platforms, unmanned systems, and novel operational constructs in the maritime littorals.

For the Coast Guard, however, the promise of force design lies less in exotic hardware than in finally aligning its unique blend of authorities, missions, and international partnerships with a realistic assessment of what the nation expects it to do.

This chapter asks whether the force design discourse meaningfully incorporates the Coast Guard as a proactive shaper of the maritime security environment, a white fleet central to day-to-day competition, or whether the service remains an afterthought, invoked rhetorically but left structurally under-resourced and institutionally under-integrated.

The final chapter, "Final Thoughts," is intentionally not a neat summation or a policy blueprint. The story traced in this book does not admit easy endings.

Instead, these closing reflections return to the human and institutional realities encountered throughout: the district commanders juggling impossible trade-offs, the crews sustaining over-worked cutters and aircraft, the acquisition leaders trying to extract coherence from fragmented budget deals, and the senior officers who understand that the Coast Guard's comparative advantage is its ability to operate in the blurred spaces between war and peace, domestic and international, enforcement and engagement.

It is in those spaces that the United States will increasingly compete, deter, and, when necessary, respond. The question is whether national strategy and resource decisions will finally recognize that fact, or continue to treat the Coast Guard as a convenient but expendable margin of safety.

Section VIII is therefore both retrospective and forward-looking. It argues that the Coast Guard's persistent under-resourcing is not merely a budgetary oversight, but a symptom of a deeper confusion about how the United States conceives of security in the maritime domain.

If we continue to frame security primarily in terms of high-end war-fighting, we will keep mis-valuing a service whose daily work lies in deterring, shaping, and managing the messy spectrum of activity below that threshold.

If, on the other hand, we take seriously the realities of chaos management, overlapping crises, gray-zone pressure, and systemic shocks to global maritime commerce, then the Coast Guard emerges not as a peripheral player but as a central instrument of national power.

The chapters that follow do not claim to resolve this tension, but they do insist that the choices can no longer be deferred. The crossroads is here.

What remains to be decided is whether the nation will continue to rely on an "always ready" Coast Guard while leaving it persistently under-resourced, or whether it will finally align missions, means, and expectations in a way worthy of the people who serve in this small but strategically vital service.

# 1

# THE COAST GUARD AT A STRATEGIC CROSSROADS: A 2011 PERSPECTIVE

I WROTE a report in 2011 which focused on shaping a narrative which would drive broader policy understanding of the USCG's core missions and operational concepts in a changing strategic environment.

Rather than including the 20,000 word report, I am including a summary of the argument for it provides a perspective from the period when I was doing significant interviewing of USCG leadership about where they found themselves in the strategic landscape.

IN AUGUST 2011, the United States Coast Guard stood at a pivotal moment in its evolution as a national security institution. A decade after the September 11 attacks fundamentally transformed American homeland security architecture, the service confronted profound questions about its identity, mission, and future trajectory.

- Would the Coast Guard emerge as a key participant in a global maritime trade safety and security system,

> assuming a leadership role in shaping international approaches to maritime governance?
> - Or would it find itself circumscribed toward more inward-looking responsibilities, focused narrowly on ports, inland waterways, and domestic security tasks?

This fundamental choice represented more than bureaucratic positioning. It would determine whether the Coast Guard evolved as a robust national security force with global reach or whether it would shrink within the Department of Homeland Security, watching its Title 10 military roles atrophy.

Understanding the strategic narrative surrounding this crossroads requires examining how the Coast Guard's placement within DHS, the evolution of its operational concepts, and the pressures of resource constraints were reshaping the service's ability to execute its multifaceted mission set.

## The Post-9/11 Strategic Realignment and DHS's Shifting Priorities

The Coast Guard's integration into the newly created Department of Homeland Security in 2003 represented a policy response to the September 11 terrorist attacks. Congress, recognizing the potential risks of subordinating the Coast Guard's diverse missions to a homeland security-centric department, explicitly mandated that agencies integrated into DHS should not have their functions diminished by the department's focus. The legislation establishing DHS outlined seven statutory missions, three of which directly related to what Coast Guard professionals termed the "Away Game"—operations conducted far from American shores to advance national interests.

Yet by 2011, a troubling divergence had emerged between congressional intent and departmental reality. When the Department of Homeland Security was established, its statutory missions included preventing terrorist attacks, reducing vulnerability to terrorism, minimizing damage from attacks, ensuring functions of transferred agencies remained robust, protecting economic security, and monitoring connections between drug trafficking and terrorism.

However, by 2011, DHS had publicly reframed its primary missions around five core areas: preventing terrorism and enhancing security, securing and managing borders, enforcing and administering immigration laws, safeguarding and securing cyberspace, and ensuring resilience to disasters.

This shift was not merely semantic. The three Away Game missions that Congress had mandated in 2003, maintaining functions of transferred entities, protecting economic security, and addressing the nexus between drug trafficking and terrorism, had been systematically de-emphasized in DHS's public presentation of itself.

This reorientation threatened to fundamentally undermine the Coast Guard's role in the international maritime domain, potentially transforming a service with global responsibilities into one confined to domestic waters and border security.

## The Resource Reality:
## Beyond "Always Ready"

The Coast Guard's motto, "Semper Paratus"—Always Ready— embodied the service's ethos and commitment. Yet by 2011, this aspirational declaration confronted hard operational realities. The service faced what could be termed "mission stress" or a growing probability of mission failure, mission shortfall, or mission avoidance resulting from the gap between assigned responsibilities and available resources.

Understanding this resource challenge required abandoning simplistic notions of Coast Guard operations. No platform works alone. This principle, fundamental to modern Coast Guard operations, needed to be driven into public debate. Mission success depends on several platforms working together, effectively connected not only within the Coast Guard but across the commercial, law enforcement, intelligence, and defense communities. Gaps in platform availability translate directly into gaps in operational readiness.

The concept of being "always ready" in a resource-constrained environment must be understood as the ability to surge to problems and crises rather than the capacity to meet all mission requirements with diminishing assets all the time. This distinction was crucial. The Coast Guard could maintain surge capability for emergencies while acknowledging that sustained, comprehensive mission coverage across all statutory responsibilities simultaneously had become increasingly untenable.

Capabilities shape success, not merely the courage of the Coast Guard's personnel. The service could be conceptualized as a set of building blocks that must work together for mission success. Absent core building blocks, the inevitable results would be mission avoidance, gaps in mission performance, or outright mission failure. This framework provided a more honest assessment of operational capacity than rhetoric about perpetual readiness regardless of resource levels.

## Recasting the Strategic Narrative: Missions, Roles, and Concepts of Operations

Effectively presenting the Coast Guard's value proposition to Congress, the executive branch, and the American public required moving beyond bureaucratic categories. The service's eleven statutory missions and six program areas, while important for internal organization, proved cumbersome for external communication. Instead, focusing on three broad roles provided a more accessible framework for explaining what the Coast Guard does and why it matters:

- Maritime Safety — protecting lives at sea through search and rescue, vessel inspections, and aids to navigation.

- Maritime Security — defending against threats through port security, drug interdiction, migrant interdiction, and defense operations.
- Maritime Stewardship — protecting ocean resources through fisheries enforcement, pollution response, and environmental protection.

This tripartite framework better captured the service's comprehensive mission set while remaining comprehensible to audiences unfamiliar with Coast Guard organizational structures.

## The Deepwater Legacy: From Isolated Assets to Integrated Systems

The Deepwater program, despite controversies surrounding specific acquisitions, represented a fundamental shift in how the Coast Guard conceptualized recapitalization. Rather than simply replacing aging vessels and aircraft on a one-for-one basis, Deepwater pursued a system of interconnected capabilities that leveraged Command, Control, Communications, Computers, Intelligence, Surveillance, and Reconnaissance (C4ISR) to achieve greater efficiencies.

This approach to recapitalization deliberately reduced the numbers of physical assets while promising to achieve greater effectiveness from remaining, modernized platforms. Critics focused intensely on problems with specific acquisitions, particularly the 123-foot patrol boats and, to some extent, the National Security Cutter. The Coast Guard responded by creating a new acquisition directorate to subsume and reform Deepwater's processes.

Yet the underlying strategic logic of system integration remained sound. The Coast Guard needed to operate as a networked force, with sensors, platforms, and command systems sharing information across service boundaries and with partner agencies. This vision aligned with broader Department of Defense transformation initiatives and reflected the realities of maritime domain awareness in an interconnected world.

The challenge was not the concept itself but rather execution, oversight, and the political sustainability of a long-term recapitalization strategy in an era of fiscal constraint.

## Building International Capacity: The North Pacific Coast Guard Forum

The Coast Guard's unique positioning as a military service with law enforcement authorities provided distinctive advantages in international engagement. This was exemplified by the North Pacific Coast Guard Forum, which brought together the United States, Russia, Canada, China, Japan, and South Korea to address shared maritime challenges.

Unlike military-to-military engagements, which could raise sovereignty and strategic concerns, coast guard cooperation focused on practical operational challenges: illegal fishing, search and rescue coordination, environmental protection, and information sharing. The Russians developed and hosted the North Pacific Coast Guard Automated System, a certificate-based encrypted platform that enabled daily information exchange about vessels operating in the Bering Sea.

This forum structure provided tools for addressing emerging challenges, particularly in the Arctic. As climate change opened new shipping routes through Arctic waters, questions of search and rescue coordination, environmental response, and traffic management became increasingly urgent.

- How would Russian, Canadian, and American responders coordinate during a major disaster involving a cruise ship with thousands of passengers in remote Arctic waters?

- How would nations address a major oil spill in this fragile environment?

The North Pacific Coast Guard Forum provided an existing framework for developing cooperative solutions.

Significantly, the Coast Guard received no dedicated funding for this strategic engagement despite leading the U.S. role in the forum. This exemplified a broader pattern: the service was expected to maintain international partnerships and global engagement while resources increasingly focused on domestic homeland security priorities. The State Department, stretched thin, generally did not participate directly in forum meetings, though it was kept informed of activities.

The Coast Guard's access to China through this framework exceeded that of most other U.S. government agencies. Because the service was viewed as a law enforcement, search and rescue, and environmental organization rather than a military force, it could engage Chinese counterparts in ways that might be difficult for the Navy or other defense agencies. This represented a significant strategic asset for the United States in managing Pacific relationships, yet one that remained undervalued and under-resourced.

## The Away Game versus Home Game: A Defining Tension

The strategic choice facing the Coast Guard in 2011 could be framed as a fundamental tension between two operational orientations. The Away Game encompassed operations conducted far from American shores: counter-drug patrols in the Eastern Pacific and Caribbean, fisheries enforcement in distant waters, international capacity building, and support to naval operations. The Home Game focused on port security, waterway management, coastal search and rescue, and border security.

Both mission sets were legitimate and important. The question was one of balance and emphasis. A Coast Guard oriented primarily toward the Home Game would find itself increasingly confined to supporting other DHS agencies' border security priorities, with its global engagement and naval cooperation roles steadily diminishing. Budgetary pressures and DHS's reorientation toward border security and domestic resilience would accelerate this trajectory.

Conversely, a Coast Guard that maintained robust Away Game capabilities could position itself as essential to multiple constituencies: the Navy for maritime security operations, the State Department for international engagement, regional combatant commanders for theater security cooperation, and law enforcement agencies for transnational threat networks.

This positioning required articulating clear operational concepts that demonstrated how Coast Guard capabilities enabled whole-of-government approaches to maritime challenges.

The risk of allowing the Away Game to atrophy extended beyond the Coast Guard itself. Maritime security challenges, from piracy to illegal fishing to drug trafficking to proliferation of weapons of mass destruction, required persistent presence and international cooperation. If the United States Coast Guard withdrew from global maritime governance, other nations would fill the vacuum, potentially in ways less favorable to American interests. The Coast Guard's unique authorities, capabilities, and international relationships represented strategic assets that could not easily be reconstituted if allowed to degrade.

## Communicating Complexity: Concepts of Operations as Explanatory Frameworks

Securing support for Coast Guard modernization and mission execution required more than listing statutory authorities or reciting bureaucratic mission statements. Stakeholders needed to understand how the Coast Guard operates, the concepts of operations (CONOPS) that translate missions into action.

Effective CONOPS communication involved several elements.

- First, demonstrating how platforms work together rather than in isolation. A National Security Cutter deploying to the Eastern Pacific for counter-drug operations does not work alone. It coordinates with maritime patrol aircraft, small boats, shore-based command centers, international partners, and intelligence agencies. This network of capabilities enables the mission; the cutter itself is one node in a larger system.
- Second, showing how Coast Guard capabilities enable other agencies and services to accomplish their missions. When the Coast Guard provides maritime domain awareness, it enhances Navy operations, supports Customs and Border Protection, assists the Drug Enforcement Administration, and provides information to intelligence community analysts. This interagency enabling role represented a core value proposition but one that required clear articulation.
- Third, connecting capabilities to specific scenarios that resonate with stakeholders. Abstract discussions of multi-mission platforms meant less to congressional appropriators than concrete examples: a Coast Guard cutter conducting counter-drug operations that shifts to search and rescue when a vessel in distress is detected, then provides disaster response when a hurricane threatens coastal communities. These real-world mission transitions illustrated the flexibility and efficiency of Coast Guard forces in ways that organizational charts could not.
- Finally, addressing resource constraints honestly rather than maintaining fictions about unlimited readiness. Acknowledging that gaps in capabilities create gaps in coverage, that the service must make strategic choices

about where to accept risk, and that recapitalization represents investment rather than expense, these messages, delivered candidly, could build credibility with stakeholders who understood budget realities.

## Looking Forward: Strategic Choices and Their Consequences

The strategic crossroads facing the Coast Guard in 2011 demanded choices that would shape the service for decades. Would the United States invest in the Coast Guard as a global maritime security force, or would it accept a diminished service focused narrowly on domestic responsibilities? The answer depended partly on resource allocation, but more fundamentally on strategic vision.

The case for sustaining the Coast Guard's global role rested on several foundations.

- First, maritime security challenges were inherently transnational, requiring persistent presence, international cooperation, and whole-of-government approaches.
- Second, the Coast Guard's unique authorities and culture provided capabilities that pure military forces or pure law enforcement agencies could not replicate.
- Third, American maritime leadership required sustained investment in the institutions and capabilities that enable that leadership.

The risk of the alternative path was equally clear. A Coast Guard focused primarily on border security and port protection would struggle to maintain its distinctive multi-mission culture. Its most capable personnel might seek opportunities in other services or agencies. Its international partnerships would atrophy. Its ability to support naval operations would decline.

And when the nation faced the next major maritime crisis,

whether natural disaster, environmental catastrophe, security threat, or humanitarian emergency, the Coast Guard would find itself less prepared to respond.

The Deepwater program's challenges had demonstrated the difficulty of sustaining long-term recapitalization strategies in turbulent budget environments. Yet the underlying need remained unchanged: the Coast Guard's fleet and aircraft inventory was aging, and modernization could not be indefinitely deferred.

The question was whether recapitalization would proceed according to a coherent strategy aligned with an expansive vision of Coast Guard missions, or whether it would become an exercise in minimizing capability losses while managing decline.

## Conclusion: The Strategic Narrative That Matters

In 2011, the Coast Guard needed to reclaim and reinvigorate its strategic narrative. This narrative had to acknowledge resource constraints while articulating an ambitious vision for the service's role in national security and maritime governance. It had to explain not just what the Coast Guard does but why those missions matter to American security, prosperity, and global leadership.

The three broad roles, maritime safety, maritime security, and maritime stewardship, provided an organizing framework for this narrative. Concepts of operations that showed how platforms work together, how the service enables interagency operations, and how Coast Guard capabilities address real-world scenarios would make the narrative concrete and compelling.

Most importantly, the Coast Guard needed to assert its value as a global maritime force rather than accepting relegation to purely domestic roles. This required pushing back against DHS's de-emphasis of the Away Game, demonstrating the service's unique contributions to international partnerships, and making the case that

American maritime leadership requires sustained investment in Coast Guard capabilities.

The strategic choice was stark: evolve as a cornerstone of American maritime power with global reach, or accept a diminished role as another agency within the homeland security bureaucracy. The Coast Guard's history, capabilities, and culture argued for the former path. Whether the nation would provide the resources and strategic support to make that vision sustainable remained the defining question of the early 2010s.

For those who understood the Coast Guard's unique place in American national security, a military service with law enforcement powers, a homeland security agency with global reach, a regulatory force with operational capabilities, the answer seemed clear.

The United States needed its Coast Guard to be always ready not just to surge to crises at home, but to maintain persistent presence in distant waters, build partnerships with foreign counterparts, and contribute to maritime security wherever American interests were at stake.

Making this case effectively, with clarity and conviction, represents an ongoing challenge for USCG.

# 2

# THE HOPE OF FORCE DESIGN 2028

THE USCG of 2016 was seriously underfunded, especially for Pacific operations, despite rising geopolitical tensions and expanded mission sets driven by policies such as the Obama-era Pacific pivot. Throughout the Deepwater period and leading up to 2015, the Service grappled with inadequate capital budgets, aging vessels and aircraft, and deferred infrastructure maintenance. Staffing shortages and long-deferred modernization meant core missions such as Arctic presence, drug and migrant interdiction, and search and rescue were challenged by operational limitations and asset availability.

The USCG of 2026 is now receiving that higher investment and then some. On July 4, 2025, President Trump signed the One Big Beautiful Bill Act (OBBBA), providing the Coast Guard with nearly $25 billion in new resources, the largest single allocation in the service's 235-year history.[*] This funding, separate from the Coast Guard's annual budget, is designed to address the service's crumbling

---

[*] One Big Beautiful Bill Act (H.R. 1/P.L. 119-21), signed July 4, 2025. See: Maritime Executive, "One Big Beautiful Bill Hands U.S. Coast Guard Billions for Shipbuilding," July 10, 2025. https://maritime-executive.com/article/one-big-beauti ful-bill-hands-u-s-coast-guard-billions-for-shipbuilding

infrastructure while investing in a new generation of technology and assets.

The Coast Guard Authorization Act of 2025 and the OBBBA have together prioritized fleet renewal and the onboarding of advanced technologies and platforms to address inaccessible and contested maritime regions, including the Arctic and the Indo-Pacific.

The Service achieved record accessions in Fiscal Year 2025, bringing in 5,204 new active-duty members, the highest since 1991, alongside 777 reservists, exceeding both its active-duty and reserve goals.[*] These gains represent a sharp reversal from four consecutive years of missed recruiting targets that had left the service approximately 4,800 personnel short.

Despite this progress, the Coast Guard continues to face structural challenges:

- Workforce Strain: Historic personnel shortages have only recently begun to ease, but retention and recruitment remain ongoing concerns, especially for specialized roles in cyber and technical fields.
- Cyber Threats: Maritime cybersecurity presents a frontline challenge. The USCG has stood up cyber protection teams and continues to face persistent digital threats, with gaps in overall cyber staffing identified in recent GAO reports.
- Mission Expansion and Complexity: Demands on the Coast Guard, ranging from Arctic operations, environmental enforcement, and mass migration to national-level border security, have increased in both scale and complexity.

---

[*] U.S. Coast Guard, Force Design 2028 Initial Update, January 15, 2026. https://media.defense.gov/2026/Jan/15/2003856852/-1/-1/0/FD2028%20INITIAL%20UPDATE%20REPORT_20260114.PDF

- Technology and Innovation: Rapid advances in robotics, autonomous systems, next-generation sensors, and communications are being integrated, but merging new systems into legacy fleets and establishing best practices remains a work in progress.

The evolution from the under-resourced and strategically overlooked USCG of 2015 to the modernizing, globally engaged USCG of 2025–2026 marks a dramatic transformation but also highlights the enduring need for agile adaptation and sustained support to meet both new and legacy missions.

Strategic guidance emphasizes deepening commitment to recapitalization, technology innovation, and workforce growth through the Force Design 2028 blueprint. Senior leadership and official documents also highlight the necessity for ongoing congressional oversight, transparent management, and future-proofing to ensure these advancements become a durable foundation rather than a one-time boost.

## U.S. Coast Guard Force Design 2028

The United States Coast Guard stands at a critical juncture in its history. After decades of underinvestment, neglect, and strategic drift, the service faces a readiness crisis where it can no longer reliably protect the American people and the homeland. In response to this existential threat, the Coast Guard unveiled Force Design 2028 (FD2028) in May 2025, a bold transformation initiative that represents the most comprehensive restructuring of the service since World War II.

The roots of FD2028 lie in a multi-year personnel and readiness crisis that crippled Coast Guard operations. The service had missed its recruitment goals for four consecutive years, leaving it approximately 4,800 members short of its target total force. The consequences were severe: in 2024 the Coast Guard was forced to consolidate 23 small boat stations, pause operations at six stations altogether, place three 210-foot Medium Endurance Cutters in layup, and sideline seven 87-foot Patrol Boats pending reactivation.

Beyond personnel shortages, the service's cutters, boats, aircraft, information technology, and shore stations were on the verge of collapse due to long-term lack of maintenance, while efforts to replace aging assets were underfunded and behind schedule.

FD2028 is structured around four interconnected campaigns: People, Organization, Technology, and Contracting & Acquisitions. Each addresses critical deficiencies while positioning the Coast Guard for future challenges.

## People: Growing and Empowering the Workforce

The Coast Guard will grow its military workforce by at least 15,000 members by the end of Fiscal Year 2028 to restore readiness, operate a growing fleet, and deploy new capabilities. This represents a massive expansion for a service that had struggled to meet even baseline recruiting goals. As of FY2025, the momentum has shifted decisively: the service exceeded its active-duty enlisted recruiting target for the first time since 2007, accessing 5,204 members at 121% of goal.[2]

The growth strategy includes modernizing accessions for both Active Duty and Reserve components, investing heavily in recruiting incentives and marketing, and expanding officer development through the Coast Guard Academy, Officer Candidate School, and direct commissioning programs. The service is fully aligning with the President's Executive Order on "Restoring America's Fighting Force" to focus on selection and promotion by merit, and is instituting a physical fitness test for all military members.

The Reserve component receives renewed emphasis, with a focus on preparation for full-scale mobilization in a time of war, national emergency, or major contingency.

## Organization: Streamlining for Speed and Effectiveness

The organizational transformation represents perhaps the most radical element of FD2028. The initiative calls for establishing a Coast Guard Service Secretary within the Department of Homeland Security, legislatively authorized, nominated by the President, and Senate confirmed, with authorities comparable to the Secretaries of other military services. This is intended to correct a historical institutional disadvantage that has contributed to the service's neglect.

Implementation has moved rapidly. The Coast Guard has reorganized 68% of its headquarters staff, created a Chief of Staff and two new Deputy Commandant positions, and eliminated 14 flag officer positions to streamline decision-making.[5] The acquisition structure has been completely overhauled with five Program Executive Offices — Surface, Air, C5I (Command, Control, Communications, Computers, Cyber, and Intelligence), Shore, and a Robotics and Autonomous Systems business line.

The organizational changes extend to geographic alignment as well. As part of FD2028, the Coast Guard renamed the former District 17 as the U.S. Coast Guard Arctic District, aligning the district's title with its area of responsibility and enhancing interagency coordination across Alaska, the North Pacific, Bering Sea, and Arctic Ocean.[6] The change is not merely cosmetic: it signals a sustained institutional commitment to the High North as a primary operational theater.

Deployable Specialized Forces will be functionally aligned under a flag officer reporting to a single Area commander, ensuring full integration under one operational commander. Coast Guard Cyber Command is also being restructured to better address threats in the cyber and space domains, aligning with other military services.[*]

---

[*] ExecutiveGov, "Coast Guard Force Design 2028 Update," January 22, 2026. https://www.executivegov.com/articles/coast-guard-force-design-2028-update

# Technology: Leapfrogging Into the Future

The technology campaign acknowledges that the service is not equipped to deliver and use today's technology needed to effectively execute missions, and is unprepared to harness advanced technology already upon it.

A centerpiece initiative is Coastal Sentinel, a next-generation maritime surveillance capability that will develop a robust and integrated sensor network collecting, processing, and combining real-time data across platforms and systems while leveraging artificial intelligence. This system aims to deliver unprecedented identification and warning of threats along America's borders and maritime approaches.

The robotics and autonomous systems push represents what many consider the most transformational element of FD2028. The Coast Guard has established a Robotics and Autonomous Systems Program Executive Office and announced nearly $350 million in investment to expand these capabilities, including 16 VideoRay Defender remotely operated vehicles, six Qinetiq Squad Packable Utility Robots and 12 mini-SPURs, and 125 SkyDio X10D short-range unmanned aircraft systems. The RAS PEO reached initial operating capability in August 2025 and is now integrating autonomous systems directly into border security and counter-narcotics operations.

Additional technology initiatives include modernizing human resources information systems to support workforce growth, implementing conditions-based maintenance through improved logistics systems, and the establishment of a Rapid Response Prototype Team to quickly identify and deliver cutting-edge capabilities to operators.

## Contracting and Acquisitions: Speed Over Bureaucracy

Consistent with the President's Executive Order on "Modernizing Defense Acquisitions and Spurring Innovation in the Defense Industrial Base," the Coast Guard is reforming its acquisition processes to successfully deliver ready, critical capabilities while managing risk.

Key reforms include establishing a Senior Procurement Executive within the Coast Guard Secretariat, outsourcing procurement activity to increase efficiency by leveraging other government agencies, and eliminating consensus-based decision-making in favor of empowering Program Executive Officers and program managers. The service will incentivize acquiring capabilities within scheduled cost and performance and hold personnel accountable for failing to deliver on-time and on-budget.

## Arctic Recapitalization: From Promise to Contract

The most visible material manifestation of FD2028's ambitions has been in Arctic icebreaking. The chapter's original text noted vaguely the promise of "new icebreakers" alongside references to delayed Polar Security Cutters. As of early 2026, that picture is dramatically clearer and more concrete.

The One Big Beautiful Bill Act provided the Coast Guard with nearly $9 billion specifically for icebreaking recapitalization: $4.3 billion for up to three new heavy Polar Security Cutters, $3.5 billion for medium Arctic Security Cutters (ASCs), and $816 million for additional light and medium icebreaking vessels.[*]

On December 26, 2025, the Coast Guard awarded contracts for construction of up to six ASCs. Rauma Marine Constructions of Finland was awarded a contract to build up to two ASCs with first delivery in 2028; Bollinger Shipyards of Lockport, Louisiana was awarded a contract for up to four, with first delivery in 2029.[8] A U.S.-Finland Memorandum of Understanding signed in October 2025 by President Trump and Finnish President Alexander Stubb formalized

---

[*]   Ted Stevens Arctic Center, "Historic U.S. Investment in Arctic Security," July 31, 2025. https://tedstevensarcticcenter.org/historic-us-investment-in-arctic-security/

the partnership, authorizing construction of four ASCs abroad to address "urgent national security needs in the Arctic region."[*]

On the heavy icebreaker side, the Polar Security Cutter program now has a name for its lead ship: the USCGC Polar Sentinel, currently under construction in Mississippi, with expected delivery by 2030.[†] The recently commissioned USCGC Storis, a converted commercial Arctic support ship acquired in 2024, is providing interim icebreaking capability from its temporary homeport in Seattle while the new fleet comes online.[‡]

Russia operates 41 icebreakers, including nuclear-powered vessels; China has built three and declared itself a "near Arctic state." The United States, which currently has three icebreakers, has signed the ICE Pact with Canada and Finland, committing to share designs and shipyard capacity toward a dramatically expanded polar fleet. The Coast Guard's Arctic District now exists explicitly to manage the operational demands this competition will generate.

## Implementation and Accountability: Early Results

FD2028 is not merely aspirational. The Commandant delivered an execution plan within 30 days of the initiative's launch, and the Coast Guard has been providing semiannual updates on implementation progress. The January 15, 2026 FD2028 Initial Update released on the same day Adm. Lunday was formally sworn in as

---

[*]   DHS, "DHS Celebrates Purchase of New Coast Guard Icebreakers as President Trump Signs Deal with Finland," October 10, 2025. https://www.dhs.gov/news/2025/10/10/dhs-celebrates-purchase-new-coast-guard-icebreakers-president-trump-signs-deal

[†]   Alaska Beacon, "As the Arctic Heats Up, the U.S. Coast Guard's Icebreaker Fleet is Preparing for Boom Times," November 24, 2025. https://alaskabeacon.com/2025/11/24/as-the-arctic-heats-up-the-u-s-coast-guards-icebreaker-fleet-is-preparing-for-boom-times/

[‡]   Alaska Beacon, "U.S. Coast Guard Adds Icebreaker to Fleet for First Time in 25 Years," August 10, 2025. https://alaskabeacon.com

Commandant documented measurable outcomes across all four pillars.*

On July 1, 2025, the Coast Guard dissolved the Deputy Commandant for Mission Support overhead structure and elevated both Personnel Readiness and Materiel Readiness functions to separate Deputy Commandants, accelerating accountability and decision-making. The Robotics and Autonomous Systems PEO reached initial operating capability in August 2025.

Three named operational campaigns illustrate the translated impact of FD2028 reforms:

- Operation Border Trident, launched March 2025, targeted transnational terrorist and criminal organizations in the California Coastal Region, increasing interdictions by 44% compared to FY24.
- Operation Pacific Viper, launched August 2025, doubled the Coast Guard's most capable assets in the Eastern Pacific. In just over four months, crews seized more than 170,000 pounds of illegal narcotics — averaging approximately 1,600 pounds of cocaine interdicted per day.
- Operation River Wall, initiated October 2025, secured approximately 260 miles of the Rio Grande River, resulting in the interdiction of 47 illegal aliens and the deterrence of 237 more.†

At the strategic level, FY2025 produced the Coast Guard's best counter-narcotics year on record: nearly 510,000 pounds of cocaine seized, more than in any year in the service's history and over three

---

* USNI News, "Billions in Funding Helps Coast Guard Rapidly Implement Force Design 2028," January 15, 2026. https://news.usni.org/2026/01/15/billions-in-funding-helps-coast-guard-rapidly-implement-force-design-2028-modernization-efforts
† https://www.news.uscg.mil/Press-Releases/Article/4374170/us-coast-guard-highlights-historic-operational-successes-in-2025/

times its annual average. The service also rescued nearly 5,000 people at sea and ensured the safe transit of 1.8 billion tons of cargo, with conservative analysis showing the service generates more than $74 billion in social and economic value through cost avoidance , a six-to-one return on taxpayer investment.

The funding pipeline is moving. Of the $7.7 billion already allocated from the OBBBA, the Coast Guard has committed to obligating 75% of total OB3 funding by the end of Fiscal Year 2026, with plans to complete the process by January 2027.

## Conclusion

FD2028 represents a generational opportunity at a critical moment in history, one that will fulfill the Coast Guard's potential to meet the Nation's needs. The initiative acknowledges uncomfortable truths about the service's recent state while providing a concrete roadmap for renewal. As of early 2026, that roadmap is no longer theoretical: contracts have been signed, ships are under construction, operations are producing record results, and the organizational framework is being rebuilt from the headquarters down.

The challenge remains immense. The Coast Guard must simultaneously fix critical deficiencies, run its operations, make lasting transformational change, and grow its capacity and capability. Success will require sustained political support, adequate funding, and cultural adaptation throughout the service. But the arc of evidence from the first year of FD2028 suggests the service has moved from crisis acknowledgment to crisis response, a transition that has been overdue for over a decade.

As geopolitical competition intensifies across the Arctic and the Indo-Pacific, and as threats to America's maritime borders multiply, the Coast Guard's transformation through Force Design 2028 may prove decisive not just for the service, but for national security itself.

# Primary Sources for this Chapter

- Force Design 2028 Executive Report, May 2025 [*]
- USCG Official Force Design 2028 Website.[†]
- GAO Report: GAO-25-107224 on Coast Guard Recruiting Challenges.[‡]
- Federal Register Notice on Operational Adjustments, April 26, 2024.[§]
- ALCGFD28 003/25 - Force Design 2028 Organizational Changes, June 30, 2025.[#]

---

[*]  https://media.defense.gov/2025/May/27/2003724531/-1/-1/0/REPORT%20-%20FD28%20EXECUTIVE%20REPORT%20_1166_V14.PDF

[†]  https://www.uscg.mil/leadership/commandants-initiatives/forcedesign2028/

[‡]  https://www.gao.gov/products/gao-25-107224

[§]  https://www.federalregister.gov/documents/2024/04/26/2024-08978/operational-adjustments-resulting-from-workforce-shortages

[#]  https://content.govdelivery.com/accounts/USDHSCG/bulletins/3e762e5

3

———————————

# FINAL THOUGHTS

"ALWAYS READY, PERSISTENTLY UNDER-RESOURCED" brings together the Coast Guard's recent history, strategic imperatives, and unyielding operational ethos to articulate a central lesson for American maritime security policy.

At the heart of this narrative is the persistent mismatch between the Coast Guard's critical, multifaceted missions and the chronic shortfalls in personnel, platforms, and funding that define its institutional reality.

The Coast Guard stands as a military service with law enforcement authorities, operating seamlessly across domestic and international theaters. It anchors homeland security and serves as an American instrument of global maritime power, yet has never received the commensurate prioritization, investment, or sustained strategic attention that its expanding roles demand.

The story the book tells is not simply one of underfunded programs and aging platforms, but of strategic adaptation amid unrelenting complexity, of dedicated professionals who improvise and overcome in the face of resource constraints, shifting priorities, and mounting operational demands.

The Coast Guard's operational journey since 9/11 charts a transition from episodic crisis management to the perpetual chaos that now defines maritime security. Terrorist threats, gray zone adversaries, criminal networks, cyber vulnerabilities, and ecological disasters converge to create an environment where boundaries between law enforcement, defense, and humanitarian response are fluid.

This complexity has given rise to the concept of a "white fleet", a service that wields military-grade capabilities in the service of persistent presence, partnership-building, and resilience, rather than overwhelming force. The Deepwater modernization wave, despite its controversies, set the template for multi-mission, networked operations and established the enduring value of integrating C4ISR (Command, Control, Communications, Computers, Intelligence, Surveillance, and Reconnaissance) capability into every physical asset. The Coast Guard's evolution away from isolated platforms and toward system-wide interoperability mirrors the broader transformation of American defense, where knowledge-sharing, domain awareness, and rapid, adaptive response eclipse traditional metrics of military strength.

Yet, as the book underscores, the Coast Guard's greatest challenge is not technological or operational: it is strategic. Each administration brings its own priorities, reorienting the Coast Guard toward different mission sets without addressing the chronic resource imbalances that undercut mission execution.

This phenomenon is not a product of poor leadership or organizational failure, but rather a symptom of deeper tensions in national security and budgeting processes. Agencies that straddle multiple policy domains, defense, law enforcement, environmental protection, and economic security, struggle to find advocates, even as their relevance grows.

The book's interviews and case studies reveal a pattern: regardless of era or leadership, the service's ethos of improvisation and problem-solving remains the Coast Guard's most reliable asset.

But the dangers of relying indefinitely on professional excellence in the absence of adequate resources are acute. Mission stress, capability gaps, and unaddressed risks multiply; readiness can no longer be taken for granted or sustained by improvisation alone.

We are facing a candid reckoning with what the Coast Guard can do with current resources and what must be left undone unless sustained investment arrives.

The Coast Guard's three overarching roles, maritime safety (search and rescue, inspections, aids to navigation), maritime security (defense, interdiction, counter-terrorism), and maritime stewardship (environmental protection, fisheries enforcement), provide a more intelligible framework for communicating value to policymakers.

These roles demand not just physical assets, but fully integrated systems for persistent domain awareness, decision support, and collaborative crisis response. The institutional memory and lessons of Deepwater, Katrina, and Deepwater Horizon populate this framework with stories of innovation and discipline, but also reminders that surge capacity does not negate the need for baseline readiness and robust capability across the mission spectrum.

The Coast Guard is at a strategic crossroads in 2025 as it was when I wrote my report in 2011. The choices facing Congress and executive leadership will shape the service's future for decades:

- Will the Coast Guard be resourced and empowered as a global maritime force, indispensable in gray zone competition and capable of shaping international security architectures?
- Or will it be tethered to a diminished role, focused narrowly on domestic borders and port protection, unable to sustain its distinctive multi-mission culture, partnerships, or international engagement?

The consequences of the latter path from atrophy of expertise to erosion of partnerships and degraded support for naval operations are stark.

I have returned to the lived experience of September 11th, the transition from Cold War legacies to the bewildering new landscape of terrorism, asymmetric threats, and dislocation in security thinking.

The Coast Guard, for all its imperfections and resource struggles, remains essential: a case study in the intersection of courage, adaptability, and institutional neglect.

Policymakers must recognize the Coast Guard's enduring national security role and commit to sustained investment, strategic vision, and operational autonomy. Failure to do so will mean asking more of a service whose capacity to deliver is already stretched beyond responsible limits.

In sum, the Coast Guard's story is the story of American security in the twenty-first century: ambitious in mission, constrained by resources, continuously adapting, and always essential.

The challenge moving forward is to harmonize strategic vision with practical support, to move beyond rhetorical commitment and toward the kind of recapitalization, integration, and innovation that American interests demand in a world defined by persistent complexity, rapid change, and blurred boundaries between peace, crisis, and conflict.

# ABOUT THE AUTHOR

A leading analyst of U.S. naval and joint force transformation, Robbin Laird has devoted a substantial and evolving body of work to the United States Coast Guard, a service whose unique positioning at the intersection of military capability, law enforcement authority, and humanitarian mission has increasingly captured strategic attention.

Over decades of engagement with America's sea services, Dr. Laird has written extensively on how the USCG operates within overlapping domains of homeland security, maritime safety, environmental stewardship, and great-power competition, documenting the service's transformation from post-9/11 homeland defense emphasis toward renewed focus on strategic competition in contested waters.

His Coast Guard analysis reflects the same rigorous methodology that defines his broader defense work: extensive field research, sustained dialogue with senior leaders and operational practitioners, and careful documentation of organizational learning processes.

Rather than viewing the Coast Guard through Washington policy frameworks alone, Laird's approach emphasizes operational realities: how Cutters actually patrol, how Sectors coordinate response, how the service balances seventeen statutory missions with constrained resources, and how Coast Guard culture adapts to expanding strategic demands while preserving core traditions of service and seamanship.

Laird's Coast Guard writing particularly illuminates the service's modernization challenges and opportunities. He has examined the recapitalization efforts essential to replacing an aging fleet, analyzing programs like the National Security Cutter, Offshore Patrol Cutter, and Polar Security Cutter within broader contexts of maritime domain awareness and distributed operations.

His work documents how technological integration from enhanced C4ISR systems to unmanned capabilities is transforming Coast Guard operations while maintaining the human judgment essential to the service's diverse missions.

This technological evolution parallels themes in his broader analysis of fifth-generation warfare, where sensor fusion and networked operations create new operational possibilities across all maritime services.

The Arctic emerges as a critical focus in Laird's Coast Guard analysis, reflecting his recognition of how climate change and great-power competition are converging in the High North. He has documented the Coast Guard's unique capabilities and responsibilities in polar regions, examining icebreaker requirements, infrastructure needs, and the service's role in asserting U.S. sovereignty and presence in increasingly accessible Arctic waters. His work connects Coast Guard Arctic operations to broader strategic competition with Russia and China, analyzing how the service contributes to whole-of-government approaches in regions where traditional military-civilian boundaries blur.

Similarly, Laird has explored the Coast Guard's expanding Indo-Pacific role, documenting the service's contributions to regional security cooperation, illegal fishing enforcement, and capacity-building with allied nations. His analysis examines how Coast Guard capabilities, particularly in maritime law enforcement and capacity-building, complement Navy operations while providing diplomatic flexibility unavailable to purely military forces. This work reflects his broader understanding of how strategic competition unfolds across multiple domains simultaneously, requiring diverse capabilities and approaches.

Central to Laird's Coast Guard analysis is his recognition of the service as integral to what he terms the "resilient national maritime enterprise" or the interconnected system of naval forces, merchant marine, port infrastructure, and maritime industries essential to national security and economic prosperity. He documents how the Coast Guard's regulatory, response, and security functions underpin this enterprise, examining the service's unique ability to integrate homeland and expeditionary missions, peacetime and wartime roles, military and civilian capabilities.

Throughout his Coast Guard work, Laird emphasizes the service's distinctive culture, the ethos of "Semper Paratus" (Always Ready) that enables Coast Guardsmen to transition seamlessly between rescuing mariners, interdicting smugglers, enforcing fisheries, responding to environmental disasters, and operating alongside Navy forces in contested waters.

His analysis illuminates how this versatile, adaptive culture positions the Coast Guard as America's maritime first responders while increasingly recognizing their strategic value in great-power competition, a recognition that demands adequate resources, modernization investment, and integration into national security planning commensurate with the service's expanding responsibilities in an era of renewed maritime strategic competition.

# ARTICLES PUBLISHED BETWEEN 2002 AND 2013

## Deepwater / USCG Acquisition

Robbin F. Laird, "Capabilities-Based Procurement: The Coast Guard Leads the Way," *Sea Power* (Navy League of the United States), August 2002.

Robbin F. Laird, "Transformation and the Defense Industrial Base: A New Model," *Defense Horizons,* no. 26 (Center for Technology and National Security Policy, National Defense University, May 2003).

## USCG Roles, Strategy, and Future

Robbin F. Laird, "Maritime Security: New Requirements for Global Cooperation," *EuroFuture*, Winter 2005.

Robbin F. Laird, "The Maritime Trade Dynamic: Reshaping the US Coast Guard Role," *RUSI Defence Systems,* 2008.

Robbin F. Laird, "Coast Guard Visions," *RUSI Defence Systems* (February 2008).

Robbin F. Laird, "Beyond Deepwater: The U.S .Coast Guard at a Strategic Crossroads," *RUSI Defence Systems*, 2008.

Robbin F. Laird, "Shaping a Twenty-First-Century U.S. Coast Guard: The Key Role for Maritime Patrol Aircraft," in *Meeting the Challenge of Maritime Security*. Trade Publication, 2008.

Robbin F. Laird and Edward Timperlake, "Pivot Point: Reshaping U.S. Maritime Strategy to the Pacific," *Jane's Navy International,* April 2013.

Robbin F. Laird, "USCG and Caribbean Maritime Security," in *Meeting the Challenge of Maritime Security*. Trade Publication, 2008.

## Interviews and Platform-focused Pieces

Robbin F. Laird, "Framing a Mission Architecture: An Interview with USCG Admiral John Currier," *Military Logistics International,* 2007.

Robbin F. Laird, "The Role of C4ISR in the U.S. Coast Guard: An Interview with Rear Admiral

Robert E. Day Jr., USCG," in *Meeting the Challenge of Maritime Security*. Trade Publication, 2008.

Robbin F. Laird, "The New U.S. Coast Guard Cutter: A Chaos Management System," in *Meeting the Challenge of Maritime Security*. Trade Publication, 2008.

Robbin F. Laird, "The Role of Maritime Patrol Aircraft in a Twenty-First-Century U.S. Coast Guard," in *Meeting the Challenge of Maritime Security*. Trade Publication, 2008.

Robbin Laird, "Shaping Arctic Strategy," *Frontline Defence,* 2012.

## Search and Rescue

Robbin F. Laird, "Search and Rescue: Why the Coast Guard Needs Help," *AOL Defense,* June 15, 2011.

Robbin F. Laird, "Brother Mountain Rescue and the Evolution of USCG Aviation," in *Meeting the Challenge of Maritime Security.* Trade Publication, 2008.

AIRPOWER AND MARITIME
FORCE MODERNIZATION SERIES

The other books in this series are as follows:

- A Tiltrotor Enterprise: From Iraq to the Future
- Italy and the F-35: Shaping 21st Century Coalition-Enabled Airpower
- A Paradigm Shift in Maritime Operations: Autonomous Systems and Their Impact
- The Coming of Maritime Autonomous Systems: Empowering and Enhancing the Kill Web Force
- My Fifth-Generation Journey: 2004-2018
- A Tiltrotor Perspective: Exploring the Experience
- The F-35 and 21st Century Defence: Shaping a Way Ahead
- Remembering the B-17 and Its Role in World War II: Noirmoutier Island, France, 2013
- Training for the High End Fight: The Paradigm Shift for Combat Pilot Training

# Series Overview

The Airpower and Maritime Force Modernization Series is a coherent body of work that tracks how democratic societies are trying to reshape air and maritime power for a much more demanding era of great-power competition, crisis management, and coalition operations. From tiltrotor innovation to fifth-generation airpower, from maritime autonomous systems to pilot training and the modern U.S. Coast Guard, the series follows one central question: how do operational forces, technologies, institutions, and historical memory interact to shape real combat and security capability, not just glossy concepts.

## Framing the series

Taken together, the volumes form a longitudinal study of Western force transformation from roughly the early Iraq/Afghanistan period through the current era of multi-domain, "kill web" or "security web" enabled operations. The organizing perspective is not abstract doctrine, but the lived experience of forces, platforms, and organizations as they adapt or fail to adapt to new strategic and operational realities.

The series traverses four intertwined axes:

- The rise of new air and vertical-lift capabilities, especially tiltrotor and fifth-generation fighters, and how they change tactics, basing, sustainment, and coalition operations.
- The emergence of maritime autonomous systems and kill-web and security web concepts, and what they mean for fleet design, deterrence, and distributed operations.
- The transformation of training and human capital, especially for combat aircrew, as technology shifts the

cognitive and organizational demands on pilots and crews.

- The enduring importance of institutional culture and history, visible both in the B-17 volume's treatment of World War II memory and in the new Coast Guard book's portrait of an "always ready, persistently under-resourced" service.

The books share an insistence on connecting platforms and systems to strategy, institutions, and people. They are not program histories in isolation; they are case studies in how real organizations navigate change under pressure, with imperfect information and often inadequate resources.

## Tiltrotor innovation and operational culture

The two tiltrotor-focused volumes, *A Tiltrotor Enterprise: From Iraq to the Future* and *A Tiltrotor Perspective: Exploring the Experience,* anchor the series in the disruptive impact of the MV-22 Osprey and the broader tiltrotor enterprise. They show how a once-controversial platform matured into a central enabler of distributed operations, long-range assault, and crisis response, and how organizations and allies had to rethink concepts of employment, logistics, and risk to exploit its potential.

*A Tiltrotor Enterprise* traces the Osprey's trajectory from combat employment in Iraq and Afghanistan into the broader redesign of U.S. and allied force posture. The book highlights:

- How the combination of range, speed, and vertical lift allowed new patterns of sea-basing, special operations support, and theater maneuver.
- How Marine, joint, and allied operators used the aircraft to experiment with new forms of distributed,

"reach-deep" operations that prefigured later kill-web thinking.

- How institutional learning—MAWTS-1, squadron-level innovation, and coalition exercises—turned early experience into doctrine and training.

A *Tiltrotor Perspective* complements this by foregrounding the lived experience: squadron narratives, operational anecdotes, and the cultural journey of aviators and maintainers as they moved from skepticism to mastery. Read together, the tiltrotor books illustrate a recurring theme in the series: modernization is not just about acquiring a new platform, but about changing mental models, training pipelines, and alliance patterns to unlock what a platform can actually do in real operations.

## Fifth-generation airpower and coalition integration

*Italy and the F-35: Shaping 21st Century Coalition-Enabled Airpower* and *My Fifth-Generation Journey: 2004–2018* extend the modernization narrative into the realm of fifth-generation combat aircraft and coalition integration. They show how the F-35, as both a flying sensor network and a multinational program, has reshaped not only tactics and maintenance practices but also alliance politics, industrial relationships, and concepts of deterrence.

Italy and the F-35 focuses on a specific national case, using Italy's experience to illuminate how a medium-sized ally can leverage fifth-generation airpower to gain disproportionate strategic effect. The book follows:

- The integration of Italian F-35s into national defense and NATO missions, including air policing, expeditionary deployments, and Mediterranean security roles.
- The way the International Fighter Training Center and Italian training approaches cultivate a new kind of pilot—

> one comfortable operating in a software-defined,
> data-rich, coalition environment.

- The industrial, political, and alliance dimensions of
  Italy's F-35 participation, which make the aircraft not
  just a capability but a vehicle for deeper integration with
  partners.

*My Fifth-Generation Journey* then steps back to offer a personal and analytic narrative of the broader F-35 era from 2004 to 2018. It connects factory floors, training ranges, squadron visits, and policy debates into a single story about how fifth-generation thinking displaced legacy platform-centric assumptions.

This volume underscores a motif that recurs across the series: transformation proceeds unevenly, with pockets of excellence and areas of inertia, and much depends on whether senior leaders and institutions can translate early adopter experience into mainstream practice. Together, the F-35 books sit at the intersection of technology, coalition politics, and training, and they feed directly into later volumes on pilot training and the kill-web.

## Maritime autonomy, kill webs, security webs and distributed operations

*A Paradigm Shift in Maritime Operations: Autonomous Systems and Their Impact* and *The Coming of Maritime Autonomous Systems: Empowering and Enhancing the Kill Web Force* shift the lens to the maritime domain and the rise of uncrewed, networked systems. These books argue that autonomous systems are not an add-on to traditional fleets, but a catalyst for rethinking deterrence, presence, and warfighting concepts from the keel up.

*A Paradigm Shift in Maritime Operations* lays out the conceptual foundations:

It explains how surface, subsurface, and aerial unmanned

systems can extend sensing, complicate an adversary's targeting, and provide affordable mass in contested environments.

It stresses that simply bolting unmanned systems onto legacy command structures does little; the true shift comes when maritime forces adopt "kill web" logic, dynamic, resilient networks of sensors, shooters, and decision-nodes that can operate under attack.

*The Coming of Maritime Autonomous Systems* then dives deeper into the operational and programmatic implications. It examines specific use-cases, experimentation pathways, and industrial choices that can either accelerate or stall the transition to an autonomous-rich fleet. The argument is that MAS are central to any realistic strategy for deterrence-forward postures in the Indo-Pacific and other theaters, where geography, anti-access threats, and resource constraints demand more distributed, risk-tolerant force designs.

These maritime autonomy volumes also bridge directly to the new Coast Guard book. Both show how under-resourced or structurally constrained institutions must think creatively about platforms, crew ratios, and multi-mission employment to fulfill expanding roles in a deteriorating security environment.

## Training for the high-end fight

*Training for the High End Fight: The Paradigm Shift for Combat Pilot Training* tackles the human side of modernization, arguing that the decisive variable in fifth-generation and kill-web warfare is cognitive capacity and training architecture, not just platform performance. In a world of sensor fusion, distributed networks, and complex rules of engagement, pilots become mission commanders inside a dynamic system, not simply platform drivers.

This volume:

- Charts the move from platform-centric, sortie-count metrics to integrated live-virtual-constructive (LVC)

training environments that stress decision-making, cross-domain coordination, and coalition interoperability.
- Explores how organizations like MAWTS-1 and allied training centers have redesigned syllabi, instructor roles, and evaluation standards to reflect high-end adversaries and contested information environments.
- Shows how training and doctrine lag, or lead, the introduction of new platforms, tying back to the tiltrotor and F-35 volumes, where early operational experience demanded new training paradigms.

By placing training at the center, the book reinforces a theme running through the entire series: technology only delivers strategic effect when human capital, institutional habits, and assessment regimes change accordingly. It also prefigures the Coast Guard volume's focus on the strain placed on people and training systems when missions expand faster than resources.

## History and memory: the B-17 volume

*Remembering the B-17 and Its Role in World War II: Noirmoutier Island, France, 2013* provides a historical and memorial anchor for the series. Rather than a generic bomber history, it uses the specific story of a B-17 and its crew associated with Noirmoutier Island to connect local memory, family narratives, and community remembrance with the broader history of World War II airpower.

Within the series architecture, this book serves several functions:

- It reminds readers that the contemporary debates about platforms, training, and force structure are rooted in a tradition shaped by immense sacrifice and hard-won operational lessons.
- It shows how airpower is experienced not just by crews and commanders but also by civilians, communities, and

future generations who interpret the meaning of wartime events through memorials, ceremonies, and ongoing research.

The B-17 story sits in productive tension with the future-oriented volumes. It suggests that modernization without a sense of historical continuity risks becoming technocratic and detached from the human stakes of conflict. At the same time, it invites readers to see today's aviators and coast guardsmen as successors in a lineage of service that stretches back through earlier air campaigns.

## The modern U.S. Coast Guard: "Always Ready and Persistently Under-Resourced"

The addition of *Always Ready and Persistently Under-Resourced: The Modern US Coast Guard Story* deepens the maritime side of the series and sharpens its institutional and political edge. Where the autonomy and kill-web volumes focus on technological and conceptual shifts, this Coast Guard book foregrounds an institution that has shouldered steadily expanding missions. homeland security, Arctic presence, great-power competition, counter-drug, search and rescue, without commensurate resources or political attention.

The Coast Guard story resonates strongly with earlier work on the Legend-class National Security Cutter and the missed opportunity to leverage that platform as the basis for a more robust, frigate-like capability for the wider sea services. In the context of the series, this volume:

- Exposes the structural mismatch between strategy and resourcing, showing how an "always ready" ethos can mask chronic under-investment in hulls, people, and infrastructure.
- Highlights how the Coast Guard already operates at the intersection of domestic security, law enforcement,

gray-zone competition, and international maritime governance—precisely the contested space where great-power rivalry now plays out.
- Connects to the autonomy and kill-web arguments by suggesting that Coast Guard platforms and missions could be central testbeds and anchors for distributed maritime networks, if recapitalization and policy choices catch up with operational reality.

By documenting the modern Coast Guard's burdens, innovations, and near-missed opportunities, the book injects a note of urgency into the series. It underscores that modernization is not only about designing future forces, but also about rescuing over-stretched institutions in the here and now, before strategic shocks expose fatal gaps.

## An integrated narrative of democratic force transformation

With the Coast Guard volume added, the Airpower and Maritime Force Modernization Series can be read as an integrated narrative about how a democratic system grapples with the demands of a harsher strategic environment. The individual books range across domains, platforms, and time periods, but several cross-cutting insights emerge:

Distributed, networked operations are the common destination. Tiltrotor mobility, fifth-generation fighters, maritime autonomous systems, and Coast Guard cutters all gain maximum value when integrated into resilient, multi-node kill webs rather than isolated taskings.

Human capital and training are the decisive bottlenecks. Whether in MAWTS-1 syllabi, Italian fighter training, or Coast Guard crews juggling multi-mission demands, the question is how to cultivate people and organizational cultures capable of exploiting complex systems under stress.

Institutional and political will lag operational reality. The series repeatedly shows operators improvising at the edge, Osprey squadrons pioneering new concepts, Coast Guard crews stretching ships beyond their planned envelopes, early F-35 units building joint tactics, while budgets, acquisition systems, and political debates trail behind.

History and memory matter. The B-17 book anchors the entire project in a recognition that air and maritime power are not just technical enterprises but human stories that shape and are shaped by societies' understanding of war, sacrifice, and legitimacy.

In this sense, the series offers more than a set of discrete case studies. It provides a layered, multi-domain portrait of how Western air and maritime forces are trying, unevenly and under pressure, to become more agile, networked, and resilient without losing the human and historical foundations that give military power its deeper meaning.

## Audience and utility

For practitioners in air and maritime forces, the series offers concrete insight into how peers and allied organizations are grappling with many of the same issues: integrating new platforms, rethinking training, experimenting with autonomy, and sustaining readiness under fiscal and institutional constraints.

For policymakers and analysts, the books collectively highlight the friction points between strategy, acquisition, and operations. They illuminate where investments yield real operational leverage, tiltrotor mobility, fifth-generation integration, autonomous architectures, resilient training systems, and where under-resourced institutions like the Coast Guard risk becoming hidden single points of failure in national and allied security.

For a broader interested readership, the inclusion of historical and experiential narratives, the B-17 story, the personal fifth-generation journey, the lived experience of tiltrotor communities and Coast Guard crews, makes the series an accessible entry point into complex modernization debates.

The books show that beneath buzzwords like "multi-domain" and "kill web" are real people and real organizations trying to prepare for wars that must be deterred, managed, or, if necessary, fought under conditions more demanding than anything seen in decades.